Environmental Problems Globally

Daniel Lachmann

Environmental Problems Globally

From Perception to Reaction

PL ACADEMIC RESEARCH

Bibliographic Information published by the Deutsche Nationalbibliothek
The Deutsche Nationalbibliothek lists this publication in the Deutsche Nationalbibliografie; detailed bibliographic data is available in the internet at http://dnb.d-nb.de.

Zugl.: Köln, Univ., Diss., 2014

Library of Congress Cataloging-in-Publication Data
Names: Lachmann, Daniel, 1981-
Title: Environmental problems globally : from perception to reaction / Daniel Lachmann.
Description: Frankfurt am Main: Peter Lang, [2016] | Originally presented as the author's thesis (doctoral)—Universität zu Köln, 2014. |
Identifiers: LCCN 2016001542 | ISBN 9783631673478
Subjects: LCSH: Environmental degradation—Prevention. | Human ecology.
Classification: LCC GE140 .L33 2016 | DDC 363.7—dc23 LC record available at http://lccn.loc.gov/2016001542

D 38
ISBN 978-3-631-67347-8 (Print)
E-ISBN 978-3-653-06603-6 (E-Book)
DOI 10.3726/978-3-653-06603-6

PL Academic Research is an Imprint of Peter Lang GmbH.

Peter Lang – Frankfurt am Main · Bern · Bruxelles · New York · Oxford · Warszawa · Wien

This publication has been peer reviewed.

www.peterlang.com

Table of contents

Figure index

Table index

Introduction

The study of environmental attitudes and behavior is a multifaceted undertaking that can be and has been approached from various perspectives in different disciplines. However, the perspectives are not alone in having many angles; environmental attitudes as well as behavior are multifaceted constructs, to express it in technical terms. This might very well be true for most attitudinal issues. Therefore, studying these issues poses a challenge for the researcher in terms of coping with this complexity adequately. Research articles in peer reviewed journals, however, often do not offer sufficient space to deal with all the intricacies of environmentalism. Consequently, much of the work done in environmental sociology, psychology or behavioral economics usually provides insight into only one particular portion of the whole story.

The present study aims at an extension of the state of research on the one hand, investigating aspects of environmentalism that research has so far neglected. On the other hand this study tries to bring together explanatory approaches that might well complement each other to get a fuller understanding of what drives people to be more (or less) environmentally concerned or to engage more (or less) than others in environmentally protective activities.

As this study is practically oriented, the first part (chapters 1–2) introduces the broader framework in which the present study is embedded and will clarify the practical relevance. It will give an explanation of what environmental pollution, degradation or pollution actually mean and which role human beings adopt as socio-economic, legal or political actors with regard to the environment. Without an understanding of the complexity of environmental problems and the human-environment connection it is difficult to draw practical conclusions from this or any other study. The role that people are expected to play, i.e. that of responsible citizens, can only be normative to a certain extent. Nonetheless, the first part of this study tries to proceed as analytically as possible.

The second part (chapter 3) ties in with this previous point, starting with a discussion of the theoretical background. The constructivism perspective explicitly needs reference to the intricacy of environmental problems. This helps to form theoretical explanations about why the (social) construction of the area "the environment" may be perceived and thought of so differently by many people. The second theoretical approach consists of a rational choice perspective. It then proceeds by helping to explain how the subjectively constructed reality, as consumer, parent, employee or employer, or as an actor within the legal system, comes to make a behavioral decision. How do our perception, our social positions and world views influence a particular choice made between multiple potential forms of action? Are values, beliefs or convictions what drive us? Or instrumental utilities which we pursue? Or a combination or interplay of both of these? Do we shift our attention to certain activities while others remain ignored, irrespective of the actual environmental impact? Do we use high-, medium- or low-cost activities to express our environmental concern? Are we active at all?

These theoretical questions are then investigated in the third part (chapter 4–8). The third part might be termed 'the empirical'. This includes a description of the data used and the methodological approach pursued (chapter 4). An additional focus is put on data quality assessment, in particular on the question of whether missing values compromise the data set and if the scales that are supposed to measure environmental concern are comparable across the different countries. This also includes a short investigation of possible response pattern biases as well as how missing values are dealt with. In the third part the operationalization of the dependent and independent variables is described.

After this technical section, chapter 5 continues with the substantive empirical analysis. First, a cross-national comparison at macro level will show that there are indeed differences concerning the perception of the environment as well as the level of environmental concern and action between countries. A detailed microanalysis of six selected countries provides insights into how far the theoretical assumptions discussed in part two substantially contribute to explaining environmentalism in the respective country.

Finally, a summary of the findings pointing out similarities and differences between various aspects under the microscope as well as country differences will follow. Further research suggestions in addition to practical implications with reference to the first, second and third part will serve to complete the study.

A preliminary note on the terminology of this study has to be made before starting with the excursus on environmental problems. The term 'environmentalism' will be used throughout the study. To avoid misunderstanding, the expression should be understood neutrally. No political meaning is implied. The term is simply used to refer to both environmental attitudes and environmental behavior.

An Excursus: Environmental Problems

To understand the relationship between the environment and society completely, from perception to reaction – if this is possible at all – one has to take a look at what is actually meant when speaking about pollution and environmental problems. In everyday language it is often used in a rather vague way without any specificity. Research papers in the social sciences often do so as well. The purpose of this chapter is to provide insight into the complexity of what can be found behind the concepts of pollution and environmental problems. Especially when investigating people's attitudes and behavior toward preventing pollution, it has to be kept in mind that pollution may have different definitions, not only from country to country but also between individuals. A few questions from the 2010 module of the International Social Survey Programme (ISSP), the third wave about the environment, may introduce the issue:

International Social Survey Programme 2010:

Q9c: Modern science will solve our environmental problems with little change to our way of life.

Q12a: How willing would you be to pay much higher prices in order to protect the environment?

Q13a: It is just too difficult for someone like me to do much about the environment.

Q13e: Many of the claims about environmental threats are exaggerated.

These four questionnaire items mention environmental problems, environmental protection, uncertainty about doing something about the environment and environmental threats. Do people think about throwing away a piece of paper or spilling oil into a river when pondering environmental problems? Is reducing people's energy consumption or donating to or actively participating in environmental protection groups when environmental protection is addressed? Do people consider environmental pollution and problems to be the same? These kinds of questions can be understood in diverse ways, "because different countries face different environmental problems and interpret the question differently" (Israel and Levinson 2004, p. 10). Consequently, results from survey data using such general statements should be construed carefully.

The following sections will depict what can be understood when speaking of pollution and environmental problems. This is necessary to comprehend the problems human societies face when trying to protect the environment. It might be an encouragement or motivation for further survey designs to go into further detail when addressing the issue of environmental pollution and protection. This chapter is also an aid for readers to get a better understanding of the issue which can be kept in mind and recalled when necessary throughout the reading of this study.

What is environmental pollution?

Pollution is "the introduction into the environment by humans of substances that are harmful or poisonous to people or ecosystems" (Perk 2006, p. 4). This anthropocentric definition can be complemented in a more general way: "a pollutant is a chemical out of place" (Hill 2004, p. 3). Accordingly, pollution does not only have man-made sources. Imagine the ash, chlorine, sulfur dioxide etc. emitted in a volcanic eruption. However, when speaking about pollution or preventing pollution, usually, anthropogenic pollution is addressed.

The following information is taken from Hill (2010) "Understanding Environmental Pollution", chapter one. The chemicals that are out of place are repeatedly a result of human activities. They become harmful when building up to dangerous levels, i.e. with increasing concentration of a chemical in the place the usually do not belong. They are often not harmful in the place where they belong unless the dose (i.e. the amount that is emitted) becomes higher than the place can tolerate. They can also be harmful in lower doses when moving to a place they do not belong. The typical anthropogenic sources of pollution are: transportation, refineries, manufacturing facilities, production of electricity and heating, agriculture including fertilizers, mining, forestry (deforesting), construction and building, preparation of drinking water and sewage and waste disposal.

The author emphasizes that pollution is aggravated by deliberate or careless behavior as well as poor technology. Without deliberate or careless pollution it would nevertheless still exist as no process is 100% efficient. Even when breathing we emit carbon dioxide; the question is, when is a chemical a pollutant and when not.

Pollution or not?

The greenhouse gas carbon dioxide CO_2 is often referred to as a major pollutant contributing to global warming. CO_2 per se is not a pollutant. Without any CO_2 in the atmosphere the global average temperature would be approximately 35° C lower than it is today (Hill 2010, p. 170, Lomborg 2007, p. 9). Considering that the global average temperature of the 20th century was 12.1° C (U.S. National Climate Data Centre 2011), life as we know it would barely be possible with an average temperature of about 20° C below zero.

In general, it can be said, that almost anything can be a pollutant if the dose is high enough, the location of pollution is vulnerable to it or the exposure to the pollutant lasts long enough (Hill 2004, Perk 2006). "Milk, fruit juice, and sugar, for example, are generally not considered as pollutants, but if directly released into surface water they are harmful to aquatic life, since the oxidation of the organic substances depletes dissolved oxygen into the water" (Perk 2006, p. 7).

First, two different types of pollutants have to be distinguished: Primary pollutants are substances that are harmful in the form they are released, secondary pollutants become harmful as a result of a chemical process (Perk 2006, p. 5). Additionally two general types of pollution sources need discrimination (Perk 2006, p. 8): Point sources occur in a single location and their effects are direct and local. Point sources are usually easy to identify, easy to control and easy to trace back to their cause. Diffuse sources are less straightforward to identify, as they occur over a wide area. They come into effect at a distance from their cause and often originate from several causes and include several substances, some of which might not be harmful on their own. They cause systemic harm. As a consequence, diffuse sources are hard to control and tracing them back to their cause is usually barely possible, although it might be possible in some instances. Point sources have local effects while diffuse sources have systemic effects (Hill 2004, p. 26). These two types of pollution sources cannot always be distinguished as sharply as the definition does. A local oil spill will have its most adverse effects where it takes place but as oil seeps into the ground and moves it can cause systemic effects too, distant in time and place.

Hill (2004, pp. 12–13), furthermore, describes two general characteristics of pollutants: pollutants move and pollutants change form. The adverse effects of pollutants are aggravated (or alleviated) by the ability of the place into which they are released to retain or to transport them as well as the pollutants' capacity to be retained, transported or change form (Perk 2006, p. 5). Chemicals are released into air, water or soil and while causing local effects they move through or with the respective medium. Thus, toxic waste spilled into a river (point source) can severely harm the local aquatic ecosystem but as it moves with the water downstream water and shores along the whole current of the river can be harmed as well (diffuse source). Although the most adverse effects of a chemical are usually local ones (primary pollutants), distant effects can also be strong, for example when the chemicals react with other chemicals found in other places (secondary pollutants). Similar scenarios can be imagined for air and soil "transportation" of pollutants.

From pollution to an environmental problem

Environmental pollution and environmental problems are not synonyms. Pollution becomes problematic when the place or system into which it is introduced is harmed by it so that its metabolism can no longer handle the substance(s). In general, environmental problems are problems of metabolism (Huber 2001) where the system's functions are disturbed. There is a limited capacity of ecosystems, and "contemporary 'ecological limits' refer to the finite ability of the global ecosystem to provide its vital services in the face of an increasing human load" (Dunlap 2010, p. 18). Emissions from the combustion of fossil fuels are absorbed and metabolized by natural sinks like forests and oceans; other substances are degraded as a natural service of microorganisms (Hill 2004, p. 13). Gases like carbon dioxide CO_2 are for example metabolized in photosynthesis. If the dose of these gases increases it takes longer for the sinks to metabolize them; they persist in the atmosphere for a longer time and humans, plants and animals are exposed to them for a lengthier interval, which can be harmful for them. Bell et al. (2011), for instance, show that diesel vehicle emissions can stimulate and reduce the performance of plants (e.g. number of dead leaves or photosynthesis). These stimulations or reductions "can all be viewed as having potential adverse ecological impacts" (Bell *et al.* 2011, p. 1990). The longer the plants are exposed, the more harm is caused. Additionally, more chances for other chemical processes are provided when chemicals last longer in one place. Nitrogen oxide NO_x can react with hydrocarbon $C_nH_{(2n)}$ (which are both emitted by combustion of fossil fuels) catalyzed by sunlight to ozone O_3. Ozone in turn is a toxic substance[1] to humans and non-humans (Sinha and Singh 2010, p. 6); or fine particulates emitted during fossil fuel combustion could concentrate in the air and cause acid rain. An overdose of a chemical that is degraded by microorganisms or via photosynthesis can harm the organism or plant and thus reduce their natural service capacity or destroy them entirely. Then, the service is no longer available at all, which in turn has consequences for the whole ecosystem. This can be local when a pond turns over and affects the local ecosystem or global when climate warms and thus affects the entire world (cf. Hill 2010). However, there is often an accumulation of such "local-scale forcings", as labeled in the natural sciences, from which "global-scale forcings" materialize (Barnosky *et al.* 2012, p. 54). For example, many local sources of CO_2 emissions contribute to global warming. These global forcings can, of course, affect local systems in return.

Effects of pollution, degradation or destruction not only can lead to worse conditions of the ecosystem or entail direct negative health effects, as thus far

1 That a harmful chemical is not harmful in any place where it occurs can be shown by the example of ozone. In the troposphere it is a pollutant dangerous for humans. In the stratosphere it is a proctectant as it absorbs dangerous UV rays from the sun (Hill 2010, p. 122).

described, but can also cause a so-called critical transition or state shifts. These result from "bottom up (local-to-global) and top-down (global-to-local) forcings" (Barnosky *et al.* 2012, p. 57). In the "state shift theory" (Barnosky *et al.* 2012) two distinct effects can cause such a state shift (e.g. turnover of a pond). First, the condition of an ecosystem is influenced, for example by pollution, slowly over time until a threshold that is hard to anticipate is crossed and causes the ecosystem to shift abruptly from one state to another. Second, there is a 'sledgehammer' effect (Barnosky *et al.* 2012, p. 52), a particular action (the authors name the clear felling of a forest with bulldozers) which ends in a not surprising and also very abrupt state shift. After this state change "it is extremely difficult or even impossible for the system to return to its previous state" (Barnosky *et al.* 2012, p. 52).

When pollution and environmental problems are addressed in the subsequent chapters of this work, the complexity that leads from a mere chemical to an environmental problem, their point and diffuse sources, and their route from local effects to global (systemic) effects and back, has to be kept in mind. This is particularly important from the constructivist's perspective where environmental problems, despite their objective matters of fact, are perceived and evaluated differently depending on knowledge, information, social interaction or ideologies. Imagine by what extent the answers to the survey questions at the beginning of this chapter may differ when they are answered before or after reading this chapter. The complexity and the fact that much of the world's environmental pollution and the resulting problems are caused by an accumulation of consequences of human behavior is a first answer to the question that will be addressed in more detail in the next chapter: Why conduct studies on individual environmental behavior?

1. Why conduct studies on individual environmental behavior?

After having returned from our excursus on environmental pollution and problems, the sociological part of this study begins with the question: Why study individual behavior? To answer the question of why there is a need for the study of human behavior toward the environment, one has to understand the relationship between man and the environment, and also the role people play. These are sketched out in the following chapter as the basic ideas and broader framework of the present study.

Human beings are not simply in direct relationship with their environment, with nature: they are a part of it. Nature is, so to speak, man's livelihood, and earth has to be understood "as our life support system" (Steffen *et al.* 2011, p. 746). Dunlap (2002) describes the relation between humans and environment in the "New Ecological Paradigm" (NEP) or later the "New Environmental Paradigm" (Dunlap and Van Liere 2008). In short, the basic assumptions of the NEP can be summarized as follows: humans are inevitably part of the global ecosystem. Human behavior has an impact on this ecosystem and, through biological as well as physical attributes, the ecosystem impacts and restricts human action. Natural laws cannot be switched off. Quite similar definitions of the human-to-nature relationship are also formulated as *ecocentrism* (Halsey and White 2009). An additional aspect of *ecocentrism* theory is that humans possess the ability to reflect on their behavior as well as its consequences. They have the responsibility to keep their behavior (and its consequences) within the limits of earth's capacities (Halsey and White 2009, p. 36). To clarify, in this context behavior can mean anything from an individual's eating habits up to industrial mass production of basic as well as luxury consumer goods.

The impacts of human activities often turn out to be in the form of pollution, degradation or destruction. Even though intentional pollution, degradation and destruction happen, as Jacobs (1991, p. 25) writes, they usually do not occur deliberately. These un-deliberate environmental impacts are negative externalities, indirect costs and "economic effects not taken into account in the decision making in a market" (Bell 2012, p. 73). This can, of course, be transferred to non-market decisions like the use of a certain technological advice that emits pollutants.

Pollution and destruction possibly result in new, or contribute to existing, environmental problems. These are for example: desertification, soil erosion and salinization, the contamination of soil and waters, water shortages. Further problems are depletion or thinning of the ozone layer which leads to a higher amount of ultra violet rays we are exposed to, photochemical smog (formation of ozone O_3 in the troposphere), exposures to fine particulates and acid rain, species extinction and loss of biodiversity. Maybe the most prominent (and perhaps most complex) issue is human's contribution to global climate change and related extreme weather events (e.g. Coumou and Rahmstorf 2012, Hill 2004, 2010, Huber 2001, 2011). As Bergmann and Klages (2012) recently reported, even in the most remote places such as the Arctic Ocean traces of human activity can be found in form of the strongly increasing amount of waste discovered on the ocean's floor.

Environmental problems, in turn, may cause social problems. These have direct and indirect effects which have acute or long-term consequences as a result. Several possible outcomes have been discussed in literature on the subject and in public: the first broader range includes health problems like respiratory diseases, rashes, cancer or even death, poverty and hunger especially in less developed nations. An increase in climate-related natural disasters like thunder storms or hurricanes in quantity and quality constitute a second group. Finally, environmental pollution, degradation and problems may cause or augment political and armed conflict over scarce resources as well as migration to escape environmental problems (e.g. Buhaug *et al.* 2009, Hill 2004, 2010, Hsiang *et al.* 2011, Huber 2001, 2011, Jerrett *et al.* 2005, Mearns and Norton 2009, Raleigh and Jordan 2009, Ribot 2009). Generally speaking, consequences of environmental problems can be social, economic and political instabilities or disturbances which in turn augment already existing problems. However, environmental pollution, degradation and problems often are also the result of social, political and economic upheavals or injustices, especially in the less developed world. Due to poor or no access to food (people might have not enough power or opportunity to get some) children are considered economic resources, as laborers, and thus the incentive to have many children is high. This, in turn, leads to population growth and, consequently, to more pressures on land and nature (Lappé 1988, Sen 1982, see also Bell 2012 Chapter 4 for a more recent discussion, also referring to these authors).

Political measures or technological innovations are crucial in counteracting these problems. Yet, they are not sufficient and often problematic. On the one hand, political legislation and technological innovation are much more easily implemented when they are accepted and supported by the people: "Technological fixes must be designed to be consistent with attitudes" (Heberlein 2012, p. 88) and "you can't implement a structural fix [like a law or tax system] unless it's consistent with general attitudes," (Heberlein 2012, p. 137). On the other hand, a change in people's behavior itself is necessary. The best technology will be of no use unless people utilize it. Additionally, there are doubts about whether technology and politics can solve all problems (York and Clark 2010), especially those where (social) control is barely possible, e.g. in private everyday behavior, or where it would be pretty counterproductive as it would confine liberty or democratic rights too much. This is expressed in a "tension between effectiveness and legitimacy" (Gunningham 2011) concerning the implementation and enforcement of environmental laws, a consideration easy to apply to technology. Some even argue that legal regulations can only be successful when linked to "a more fundamental cultural shift in thinking about the environment" (Mares 2010, p. 289). Or to alter the perspective: legal (and also technological) changes are supposed to follow social changes.

Another critical point to mention is when an actually more efficient technology results in a higher environmental (or social) impact rather than reducing it because the better technology is used more frequently. For instance, due to lower prices, the so called Jevons paradox or rebound (Alcott 2005) comes into effect. "A more fuel-efficient car enables one to drive more" (Alcott 2005, p. 10) and overall one would use more fuel. Therefore, "addressing ecological crises requires changes in the conditions of society at large" (York and Clark 2010, p. 491), which includes technological and legal innovation as well as behavioral changes allowing less and more smart consumption and use instead of always just producing more. The argument of using less instead of producing more can also be found in Bell (2012, pp. 7–12), describing the problems of energy production stemming from both 'traditional' sources like coal, oil or nuclear power (e.g. carbon emissions, air pollution, nuclear waste) and those that come with renewable sources (e.g. sight and sound disturbance of wind turbines, disturbance of ecosystems by hydropower dams, additional land use to grow bio-fuels). The result can be what White (2011, p. 45ff) calls "paradoxical harms" where environmental protection in one aspect leads to environmental degradation in another one. Contrary to externalities these degradations are usually known or easy to anticipate. To give an example of problems arising in connection with renewable energies, in the Northern German city of Geesthacht, in Schleswig-Holstein, a large pump-storage hydropower plant soaks water from the river Elbe into a basin. When there is high energy demand the water is released into the river to produce additional electricity. Unfortunately, when water is soaked in, the fish are too. No device to protect the fish from getting caught and killed was installed (Schuh 2012). Technological environmental protection, badly implemented though, resulted in technologically induced environmental degradation as an unintended, but maybe not unexpected, consequence. Using less, in this instance energy, would mean less risky technology, i.e. where accidents or bad implementation can cause damage, or less 'bad' technology, i.e. technology directly causing harm e.g. by pollution, will be needed. Using less is only possible with behavioral changes. This statement entails two questions: Who should change their behavior? Who is responsible for change at all?

With freedom comes responsibility, which cannot simply be transferred to the government or its institutions. The people are key social, economic, political and ecological players and for that reason their behavior needs scrutiny. Behind governments or corporations there are also always people as actors. According to Nobel laureate Elinor Ostrom (2011, p. 267) it is "the day-to-day activities of individuals, families, firms, communities, and governments, especially those in the more developed world" that must be addressed regarding climate change, but also other environmental problems, be it on a local or global scale. Billions of individual stakeholders' actions have an often negative influence on the environment (Ostrom 2011, p. 269). Yet, there are chances, too: billions can reduce their impact. The opportunities and possibilities of individuals or groups of individuals to do something in response to the ecological crisis are manifold, especially concerning consumption, energy use and transportation (Dietz *et al.* 2009, Ostrom 2011). The individual is the basic actor in any economic and social process. It is the worker in a factory, the farmer or fisher 'producing' food, the manager signing international trading contracts, the politician enacting a law as well as the person who purchases a certain good or has dinner with his or her family.

To have a sustainable economy that works for people in the context of society as a whole, people's interests, views, needs and fears, etc. as well as their capabilities must not be neglected; but on the political agenda they often are (Beck 2010). Hence, a clear investigation of people's attitudes and behavioral patterns is necessary to meet these requirements and to open up the opportunities that are offered by a sustainable citizen, as the following examples will show: Dietz et al. (2009, p. 18452) find that about 20% of CO_2 emissions from households' energy use could be saved in the next ten years by simple "low-cost emissions reductions" (p. 18452). Such low-cost reductions are, to give some brief examples, heating more efficiently, using lower but still effective laundry temperatures or simple changes in day-to-day activities, e.g. unplugging appliances instead of using stand-by mode, car pooling, etc. Regarding these issues, "the most effective interventions [...] involve combinations of mass-media messages, household- and behavior-specific information, and communication through individual's social networks and communities" (Dietz *et al.* 2009, p. 18454). Other authors find that an accumulation of the small individual effects can even mount up to a 30% reduction of energy consumption by households through changing transportation patterns, e.g. using public transport instead of the car or using more modern technologies ((Vandenbergh and Steinemann 2007). Individuals are, so to speak, "a means of tackling environmental problems in Western democracies" (Barr *et al.* 2011, p. 1225).

As it is time to start doing something now, especially faced with the fact that we are "approaching a state shift in earth's biosphere" (as Barnosky et al. 2012 name their paper), besides researching for technological fixes and waiting perhaps in vain for large-scale political solutions (Ostrom 2011) it is time to investigate the factors influencing individual's environmental behavior and action. Findings from such research are meant to aid local and national policies in particular but also global policies to create general social, political, legal and structural conditions to facilitate people's behavior. This allows a better and more efficient transfer of relatively high pro-environmental attitudes into action; or makes environmentalism so easy that no such special attitudes are needed. It enables smart solutions to be found – a term usually connected with Bjørn Lomborg concerning technology, technology policies and market policies (2010 ed.) – to deal with environmental problems via social policies.

This study has so far addressed economic, social, political, or legal aspects and made reference to the role of people. Therefore, individuals can be considered as socio-economic, political and legal actors. Before turning to the theoretical approaches that help explain the individual's environmental attitudes and their behavior, this threefold perspective will be elaborated in the following chapter to give further nuances to this general framework. In this instance, the more specific role of people is likewise discussed.

2. Environment and society: socio-economic, legal and political actors

2.1 Economy, society and the socio-economic actor

Economic aspects and elements play a crucial role not only in environmental behavior studies but also concerning environmental policy and governance as well as sustainable development. A higher income, for instance, provides more monetary resources that can be spent on environmental protection but also on other consumer goods which, in turn, can lead to unintended unsustainable consumerism and mass production. "Capitalist globalization, bolstered via neo-liberal state policy, has increased the potential scope of environmentally destructive activities" (White 2009, p. 78). Little-regulated mass production and consumption on a free market with the aim of low production prices and high sales volumes endangers the environment by emitting excessive amounts of pollutants (e.g. CO_2, SO_2, fine particulates or nuclear waste) or exploiting regenerative resources faster than they can recover (e.g. fish or wood). The treadmills of production and consumption (Schnaiberg 1980) are put into operation and we are not allowed to fall off the treadmill. People are the basic actors who have to deal in some way with all these complexities as consumers.

Conversely, legal restrictions on economic behavior, i.e. consumption and production, of people, companies or countries with the purpose of a more sustainable economy might face the counter argument of hampering progress and growth and, thus, jeopardizing jobs and prosperity. Yet, controlling pollution of resource extraction, for instance, can lead to less environmental damage and to cost-effective pollution reduction or resource management strategies. However, it can also result in more unequal distribution of environmental "bads" (Ringquist 2011, p. 300) as an unintended consequence or paradoxical harm. People also have to deal with this side of the coin in their environmental reasoning.

Having sketched out these brief examples, a struggle between economy and environmental protection seems to be going on (Porter and Linde 1995, p. 98). Yet it turns out that "most conflicts between economy and ecology disappear when we switch from a short-term to a long-term perspective" (Liebe and Preisendörfer 2010, p. 154). This implies that the economy and the environment are not antagonistic but interwoven aspects of human life. Some examples might illustrate the issue:

Contrary to trade association fears, Apel (1992) found positive rather than negative economic impacts of political measures to change people's transportation behavior toward less car driving in nine European cities. Obach (2009), for instance, reports that in developed nations, restructuring in various sectors toward sustainable production due to environmental policy instead of mere profit-oriented management created new jobs rather than increasing unemployment. Typical sectors are ecotourism, clean technology production and maintenance, the energy sector, cleanup and restoration, recycling or research on environmental issues. Job losses are unusual but may nevertheless happen (Obach 2009, p. 231 ff.) and need careful consideration, too. The complexity of the job issue in relation to environmental regulation or changes in consumption patterns can be summarized as follows: it "will affect workers through the loss of jobs, the changing of jobs, and the creation of new jobs, hopefully the so-called 'Green Jobs'" (Räthzel and Uzzell 2011, p. 1215). Horbach et al. (2008) analyze the growing importance of the environmental sector in Germany and expect a positive trend for jobs, especially for firms that are innovative, employ better-educated people and concentrate on cleaner technologies instead of end-of-pipe solutions. Regulations concerning mining or forestry, for instance, usually create more jobs than are lost. The problem is that those who lose their job do not always have the opportunity to get one of the new jobs, usually due to a lack of qualification or spatial distance to the newly available positions (cf. Obach 2009). Even if the old workers find new employment, a problem of identity linked to the social position of a certain occupation might occur as "most workers see the content of their work as a central part of their life" (Räthzel and Uzzell 2011, p. 1219). Despite the positive net effect of environmental regulations on jobs, a conflict does then exist, as shown. Regulations must, therefore, be based on broader, well-considered, trans-boundary and trans-disciplinary grounds. The individual (but not in a sense of an atomized utility maximizer in a neo-classical manner), not just abstract economic or ecologic considerations, needs to be included in the debate on how to solve environmental problems. Why would a person start behaving in an environmentally friendly way when this might entail a loss of his or her job? Or, how would he or she do so when there is no time available due to the job?

The idea of an interrelationship between ecology and economy[2] (and other social sub-systems) is a crucial background thought when considering economic

---

2 In fact both words, economy and ecology, are quite similar in terms of basic meaning. 'Eco' comes from the Greek word 'oikos' which mean house/habitation. 'Logy' is derived from 'logia', 'the study of' and 'nomy' from 'nomos', 'the managing'. Ecology (or Greek

and positional factors (on both micro and macro levels) which might influence pro-environmental behavior of the public. This is true in particular when combined with the important role of the public as consumers, energy users, employees or a person in an economically precarious situation and the fact that economy and ecology can be improved simultaneously. Likewise, it is important when then drawing conclusions and making policy suggestions from the findings. "Economics is a fundamentally social process" (Bell 2012, p. 77), and people are socio-economic actors.

2.2 Environment, law and the legal actor

The destruction of the natural environment is widely accepted as a severe threat to human society in the long run and one instrument to use against this threat is environmental law. Despite the existence of environmental laws, environmental destruction takes place, often for short-term economic profits and "remains a low priority for the international enforcement community" (Environmental Investigation Agency 2008, p. 1). In this regard especially, environmental crime is an attractive and widespread business. Brack (2009, p. 487), for instance, reports that according to Interpol estimates, after illegal drug smuggling the trafficking of endangered species is the second most frequent international crime. People are often the consumers of the products of this trafficking, maybe not knowing about the illegality involved. International institutions like the G8, Interpol, the EU Environment Programme, the UN Interregional Crime and Justice Research Institute and Environmental Investigation Agency have established five areas of illegal behavior that define environmental crime. These are 1) illegal trade of endangered species (both plants and animals), 2) illegal trade of substances that deplete the ozone layer, 3) illegal transport and dumping of hazardous waste, 4) fishing that is illegal, 5) unregulated or unreported and violations of national timber logging, harvesting, transporting or trading laws (cf. Environmental Investigation Agency 2008). These activities are considered as environmentally harmful by law from a legal-procedural stance: what is harmful or not is decided by the law (Halsey and White 2009, p. 27). Environmental laws developed for fighting these kinds of deviance not only require implementation that underlines the importance of environmental protection and enables courts to use adequate (financial) penalties (Jack 2011, p. 95), they also must be enforced "effectively and efficiently" (Gunningham 2011, p. 169).

'Oikologia') means 'study of the household', and economy (Greek 'Oikonomia') 'managing of the household' (Online Etymology Dictionary).

This is a crucial "ingredient in protecting the environment and reducing environmental harm" (Tomkins 2004, p. 294). Legal enforcement requires not only detection of a deviant act but also arrest, prosecution and finally conviction of the law breaker (Akella and Cannon 2009, p. 529).

Even in developed countries, like those of the European Union, laws are better established and more resources for effective, efficient and fair enforcement are available. Any yet the EU "Member State's poor record in implementing and enforcing EU environmental law is well known" (Jack 2010, p. 74). In international environmental law the obligation to prevent trans-boundary environmental damage and to pay for it if prevention is not possible lies with the individual states (Mafessanti 2009, p. 88). The state has to identify the source of the environmental harm, be it caused by an individual or corporate actor. The state, therefore, has to enforce the laws, based on the principle 'the polluter pays' (Atapattu 2006, p. 473 ff.). The question then is "where to intervene and how to intervene" (Gunningham 2011, p. 170)?

Jack (2011, p. 74) discusses that there is more potential for enforcing environmental law in the European Commission and the European Court of Justice but that it is not fully made use of. The difficulty coming up is whether there are indeed enough resources for implementation and enforcement available, especially when Member States fail to comply with Court decisions to prevent or repair environmental damage. In this point, Jack (2011, p. 76) argues that a crucial problem is the Member State's objection to establishing an independent institution for the inspection of each case which then would be able to know how and where to intervene. Hence, a general problem is the "lack of inspection powers" of the Commission because the Commission has to adduce the relevant evidence to bring an issue before the Court. Another problem is that EU laws often differ from individual state laws which can lead to implementation problems (Umweltbundesamt (Federal Environment Agency) 2010).

The same is true and even more complicated concerning international environmental law where no institution like the EU 'surrounds' the states and thus helps to implement and enforce legal regulations. Additionally, the burden of proof lies with the Commission at EU level (Jack 2011). Then, going down to a more local level, the evidence often has to be adduced by those who are exposed to the pollution and usually have less resources available than the polluter (e.g. a firm or organization) possesses to defend itself against accusations (Messer and Shriver 2009, p. 664). On a global scale the conflict between cross-national or international environmental protection on one side and local or national interests, policies and legal regulations on the other are profound obstacles for effective and efficient law enforcement. In a case study on four developing countries (Brazil,

Mexico, Indonesia and Philippines), the major problems enforcing environmental law were, not surprisingly, more diverse than in the EU example: "Poor interagency cooperation" go hand in hand with "inadequate budgetary resources", "technical deficiencies in laws, agency policies, and procedures", "insufficient technical skills and knowledge" and a "lack of performance monitoring and adaptive management systems" (Akella and Cannon 2009, p. 548).

When keeping in mind that the adverse effects of environmentally harmful activities do not always occur in the location where the activity takes place (see the section entitled 'An excursus') the issue becomes even more complicated.

The legal-procedural perspective discussed so far is not unproblematic. The catalogue of activities which are criminal by law does not include every form of harmful behavior. Practices of corporations and individuals that might be harmful but are socially accepted or economically beneficial are nevertheless carried out. Short-term (economic) interests might inhibit the likelihood that these actions will become illegal, legally regulated or avoided as the wrong thing to do. Some of these are everyday practices of individuals like wasting water, energy or fuel or using non-recyclable products as well as activities of corporations like badly considered construction of buildings or roads, using certain chemicals like cyanide for mining or the use of unsustainable fishing techniques which in sum cause substantial environmental harm. Many legal activities often turn out to have more severe consequences than illegal activities when taking long-term effects into account (cf. Halsey and White 2009, p. 28). White (2011) provides an extensive framework of what he terms 'eco-global criminology' that goes far beyond the legal-procedural dimension of harm against the environment and its 'inhabitants'. One crucial aspect he emphasizes is that it is the harm itself that should be the center of attention and not the act of doing something illegal (White 2011, p. 22).

At this point, one could wonder what all these procedural, formal and institutional issues of environmental laws have to do with the present study and people's environmental attitudes and behavior. First of all, for efficient enforcement it is also vital that laws are accepted. At this point people come into play. Laws need legitimacy in the population, they must be enforced fairly (as well as effectively and efficiently) and need to be formulated clearly (Akella and Cannon 2009, p. 529).

In an international context and with the trans-boundary dimension of environmental crime and pollution, cooperation between organizations such as the European Commission, the European Law Enforcement Organization (Europol and other law enforcement agencies at global and local level), (Member) States, NGOs, the police and *individuals* is necessary to overcome implementation and enforcement problems, to "co-ordinate complaints" (Jack 2011, p. 79) and, thus, gather evidence more quickly and efficiently. Besides the necessity of legitimization and acceptance by the people, the people also raise the most complaints about environmental infringements (Jack 2011, p. 79). Support from the population is inevitably needed. When evidence has been found, penalties for the infringement and additional lump sum penalties for a delayed compliance can be imposed, in the name of the people.

Environmental laws not only provide one important tool to deal with environmental problems. As already stated above, individuals have been found to make the most complaints concerning environmental infringements (Jack 2011, p. 79). Research also indicates that individuals are more likely to be prosecuted than corporations if they are the perpetrators (Walters 2010, p. 313). Furthermore, because individuals' everyday practices can cause substantial environmental harm, especially in the long run, along with the fact that people's political activism can put pressure on policy and law makers (see 2.3), the consideration of environmental laws and the demand for, as well as the acceptance of, environmental laws are important factors to be considered in the context of pro-environmental attitudes and especially political pro-environmental behavior. "The content of law reflects the conditions of its society" (Clinard and Meier 2011, p. 37). "Publicly proclaimed laws and regulations" (Hirschman 1986, p. 146) can aid changing people's values and thus create behavioral standards. Well-considered and implemented regulations along with social, economic and structural changes and adjustments will be a more powerful incentive in fostering voluntary pro-environmental action than mere prohibitions. On this background, the issue of the support for laws, i.e. the individuals' acceptance of environmental laws (which is a so far quite neglected issue) at micro level will be integrated into the empirical analysis. People are legal actors.

2.3 Environmentalism, politics and the political actor

As of 1972 the Club of Rome published a report 'The Limits to Growth' (Meadows *et al.* 1972). This report provided impetus for the global upsurge of the environmental movement and the formation of its branches such as the anti-nuclear power movement or the anti-genetic-engineering movement (Huber 2011, p. 109). Environmental activism and movements are, according to Buttle (2003), crucial for environmental reform. Buttle discusses three more mechanisms in addition to environmental activism. The role of state regulations (1) has already been introduced in the previous section as well as international governance (2) which goes hand in hand with state regulations. The third mechanism, termed ecological modernization, (3) can, although not completely, be compared to the idea of social and economic changes discussed in section 2.1. Environmental activism (4) as the fourth mechanism will now be discussed. To begin, a certain role is assigned to environmental activism and movements:

> "environmental movement organizations and related coalitions of new social movements and global social movement groups shape and reshape identities and build political momentum; and political mobilization serves to place pressure on state officials and private decision makers to respond to the environmental agenda" (Buttel 2003, p. 315).

The coalition with other social movements, such as human rights or health movements shows again the social complexity of environmental problems. The working together of several movement arms – the 'eco-social' movement – makes it possible to make people aware of problems and of the crossover between various areas and problems which society faces. People sensitize other people regarding problems. Thus, one point is that environmental activism can increase environmental concern among people not engaged in activism. Besides direct pressure put on policy- and other decision-makers by, for instance, a demonstration or a petition, additional indirect pressure from the population driven by a high level of general concern for the environment and an increase in consumer power (e.g. boycotts) (Walters 2010, p. 317) will consequently lean on politicians and other stakeholders. Environmental activism is, therefore, one important driver of both social and economic changes by shaping basic environmental identities as well as promoting the fostering of state regulations and international environmental governance. Thus it directly complements the socio-economic and legal perspective on the environment: environmental activism supports people as socio-economic and legal actors. It creates "people power" as Walters (2010, p. 318) terms it.

Two basic mechanisms of recruiting new members for a social movement group have been identified. Besides recruiting new followers via existing social networks, moral 'shocks' applied in the form of "powerful condensing symbols" (Jasper and Poulsen 1995, p. 493) are used to attract strangers outside the networks. Animals in poor conditions held in small cages, sea birds covered in oil or dramatic pictures of clear felling the rain forest are stereotypes of powerful condensing symbols. Using these symbols, on the downside, one runs the risk of exaggerating issues (recall the ISSP questionnaire item "Many of the claims about environmental threats are exaggerated" presented on page five), or at least they seem to exaggerate and thus rather distract people. Transferring the issue into an emotionally confined discourse of opinions and not focused on the real world problem might be the consequence.

Again, people play the central role as it is they who go out onto the street or distribute petitions and engage in other political actions. People are political actors.

After this extensive and in part consciously digressive practical framework describing the extent to which this study is embedded in real world problems and the role people actually play, attention will now shift to part two, the core of the study. Part two comprises the theoretical perspective on how the roles of the socio-economic, political, and legal actors are approached. In this part, the state of relevant research on environmentalism is summarized as well.

3. Sociological theories on environmental attitudes and behavior

This study has so far focused on a very broadly practical and policy-oriented perspective. It has described how people act in certain ways, that there are conflicts between economic and ecological reasoning and the legal or political circumstances in which people live. Pressure not to fall off the treadmill of consumption is felt by people and yet at the same time environmental concern might be present and stimulate the individual to consider stepping off the treadmill. In this chapter the theoretical perspective will be elaborated. It provides an answer to the question of how the perspective of socio-economic, legal and political actors will be approached analytically. Two general theories provide the underlying basis guidance for the study. More specific theories on the macro level and micro levels will help to identify the concrete factors that influence environmental perception, evaluation and reasoning, and, in the end, environmental behavior. These are factors that open one to environmentalism, detain one from it or pose conflicts between both positions.

3.1 General theories of individual environmentalism

Two general theoretical perspectives of individual environmentalism will be considered as the starting point of the investigation. A constructivist perspective serves to explain the 'subjective realities' on which behavioral decisions and corresponding attitudes are based. A rational choice perspective will provide the background for the analysis of behavioral decisions themselves. Therefore, as a very basic foundation for individual decision making, the constructivist perspective will be discussed first, followed by the complementary rational-choice perspective.

3.1.1 The constructivist perspective

There is definitely much that is real about the world, about the environment, about nature itself. Human perception, in contrast, is very restricted and it is only possible to snatch a small excerpt from reality. In our everyday awareness, many things remain undetected, either because they lie out of sight, for instance, behind a mountain or behind us, or they are unperceivable by merely using our senses. No one is able to see an atom with their bare eyes, neither is it possible, at least in most instances, to recognize harmful pollutants in the air unless one lives in an area where the concentration is high enough. Besides our five senses other sources are available to 'perceive' reality. Interaction with family members, friends, and co-workers, etc. provides information exchange. Additional perception prompts come from the intake of media information or during education; or socialization processes in general. Considering social interaction and mass media consumption "each person perceives the world differently and actively creates their own meaning from events" (Burr 2003, p. 19). In the words of Faris (1937, p. 150) it is "a world of 'cultural relativity'" that we live in.

Accordingly, the concept or meaning of nature and the environment is also socially constructed. Often with the distinction that natural = good and unnatural = bad (or at least less good) (Bell 2012, pp. 202–206). But what is natural? And what is unnatural? Which things that humans do or create are natural and which are not? As already pointed out in chapter 1, humans belong to nature, and to go along with Bell (2012, p. 205):

> "anything humans do must therefore be natural, including plastic – as well as cities, automobiles, toxic waste, and even atomic bombs. Indeed, what human has ever broken a law of nature? If it were a law of nature, how could a human have broken it? By this logic, nothing could be more – or less – natural than an atomic bomb."

The concluding point would not be that natural = good and unnatural = bad. Instead the following questions are relevant: What aspects of nature and human behavior are socially constructed as good? What is defined as bad and, coming now to the topic of pollution and environmental problems, what is defined at all?

Environmental pollutants and the risks stemming from them do exist, whether we perceive them or not, regardless of whether we construct them as a problem or not, and regardless of whether we like their reality or not. According to what is stated above, environmental pollution, degradation and problems can be human made or non-human made instead of natural or unnatural. Nonetheless "they induce systematic [...] harm, generally remain invisible, are based on causal interpretations, and thus initially only exist in terms of the (scientific or anti-scientific) knowledge about them" (Beck 1992, p. 23). Environmental problems are real but their perception, interpretation and reaction toward them are "open to social definition and construction" (Beck 1992, p. 23) through the social interaction already mentioned and, maybe more importantly, through the input of mass media. Even realist critics of constructivism acknowledge "that there is a definitional or constructivist dimension to environmental problems" (Dunlap and Catton 1994, p. 20).

More or less rational decisions (see chapter 3.1.2) need to be based on what one thinks and knows and on the reality constructed from that knowledge. Then, certain action is taken or not. "People may not always be aware of the environmental consequences [...] of their behavior and therefore unknowingly perform actions that increase or decrease their energy use" (Gatersleben *et al.* 2002, p. 338). To judge a product's environmental impact is a challenging task which requires some effort (Tobler *et al.* 2011, p. 591). This argument leads to the assumption that knowledge, events, and consequences that are outside the constructed world as well as hidden, i.e. not perceived, costs and benefits cannot be taken into account when an action considered. Hence, for the analysis of environmental attitudes and behavior a constructivist perspective is an important theoretical approach to their investigation and interpretation. A perspective of 'realist constructivism'[3] is thus assumed in this study. Based on the social construction of reality, the study will suggest that environmental action should follow some rational considerations (be it value or instrumentally oriented) as will be described in the next section.

3.1.2 The rational-choice perspective

Within the framework of this study the rational-choice (RC) perspective is more closely related to behavior or action than the theory of constructivism. RC is complementary to constructivism. The question of the RC frame work is what do people do with their socially constructed reality? After a brief summary of the RC idea, three connected perspectives found in literature on environmental behavior will be discussed in more detail: subjective rationality, the low-cost hypothesis, and the attention-shifting strategy.

In the very basic idea of RC people are supposed to choose alternatives whose utility is highest and whose costs are minimized; "to act rationally, then, simply means to choose the highest-ranked element in the feasible set" (Elster 1986, p. 4). However, from a contemporary perspective with the theory of constructivism in mind it is a "rationality that is both bound to and subject to social influence" (Folmer 2009, p. 267). Even though full rationality assumes perfect knowledge, it is quite clear that usually no one has full information (e.g. Hirschman 1986, Kroneberg and Kalter 2012, Opp 1999) and hence "actors make use of limited information and apply heuristic stopping rules" (Folmer 2009, p. 268) to achieve their goals. Concerning different types of environmental action Tobler et al. (2012,

---

3 The term realist constructivism is adopted from the work of Barkin (2003) that shows that realist and constructivist world views can be compatible in the research of international relations. In the present study the realist simply refers to the fact that it is assumed that there is some objective reality as described above but action follows the subjective construction.

p. 204) interpret that their "findings might support the assumption that consumers make reasoned decisions, weighting costs and benefits of an action and choosing the option they believe to have the best balance" and an emphasis should be placed on the word 'belief'.

Several aspects such as time spending and savings, monetary expenses and revenues, appreciation and disapproval by others, a bad conscience due to violation of one's values opposed to a warm glow satisfaction by sticking to them, create costs and utilities; or, as Boudon (2003) would argue, people have reasons on which to base their action decision. Due to the variety of such reasons, Max Weber (1978, pp. 24–26) established the terms 'value-rational' and 'instrumentally rational'. In value-oriented rationality, beliefs, ideals and values and the meaning ascribed to them are drivers of action; it is not oriented toward the aim and success of ego. The intrinsic value of the action itself is in the center. An actor has reasons to think that their action is the right thing to do even though the actor may not feel any of the action's consequences. Instrumental rationality, in contrast, aims at one's own success. To achieve a certain goal or outcome is in the center of interest, "the end, the means, and the secondary results are all rationally taken into account and weighted" (Weber 1978, p. 26). The distinction between these two 'rationalities' is crucial for the analysis of environmental behavior, because both 'soft' reasons such as values (value-rational) and 'hard' reasons such as monetary expenses or health risks (instrumental-rational) play an important role as behavior's determinants, as will be discussed in chapter three.

To promote pro-environmental behavior market incentives (attenuate external conditions e.g. lower prices or better accessibility; extrinsic motivation) can be provided that lower the individual's cost (instrumentally rational). However, when considering that attitudes, subjective perceptions and norms also play an important role (value-rational), environmental morale and the way of thinking about environmental issues (environmental attitudes) in general (Frey and Stutzer 2006) must be taken into account as well promoting pro-environmental behavior (intrinsic motivation). From a rational choice perspective this means increasing the personal utility of the action such as "autonomy, competence, self-acceptance, affiliation, and […] sense of community or health" (Tabernero and Hernández 2011, p. 661). The action is perceived as the right thing to do, a "warm glow (or intrinsic) satisfaction" (Clark *et al.* 2003, p. 238) is then obtained. So, intrinsic end extrinsic incentives both seem to help foster pro-environmental behavior, or, formulated differently, to lower its costs and give good reasons. Unfortunately, intrinsic and extrinsic motivations are found to correlate negatively (Pelletier *et al.* 1997, Sherman and Smith 1984, Tabernero and Hernández 2011).

This means that with increasing intrinsic motivation a decrease in the effect of extrinsic motivation can be observed, and also the other way around.

The present study therefore assumes what Opp (1999) would term a "wide version" of rational choice in which besides egoistic preferences and material constraints also values, beliefs or norms play a role in determining an individual's action. In this study general values as well as environmental attitudes are assumed to determine the preferences and norms of action for the value-rational perspective. People can hold normative attitudes toward the environment. If they believe that protecting the environment is the right thing to do, they might value-rationally decide to act accordingly. From an instrumental rational perspective, more material aspects determine preferences. On the one hand, the perception of environmental problems as dangerous can easily be supposed to promote pro-environmental instrumental-rational action. The preference is to prevent health and safety threats. On the other hand, the individual economic condition might inhibit pro-environmental actions because more costly actions are avoided either as a result of monetary constraints or a preference for not paying for environmental protection. Due to the fact that the ISSP does not include measures that directly evaluate the individual's perceived costs and benefits, an indirect testing strategy (see Kroneberg and Kalter 2012) will be pursued. The exact procedure will be described in chapter 4.

There are still critics of the wide version of rational choice theory (see Kroneberg and Kalter 2012, Opp 1999). Concerning the present study of environmental concern and action, the wider version and the integration of the constructivist perspective is necessary because environmental issues have a material and an ideal or normative side meaning that mere utility maximization of the narrow version is not sufficient.

So far RC has illustrated why people would act in an environmentally friendly way. Three more specific approaches try to account for reasons why people do not act in this way, and why they don't do so despite their pro-environmental attitudes.

Subjective rationality

The subjective rationality hypothesis starts with the assumption that "environmental behavior is often trapped in a specific dilemma situation" (Diekmann and Preisendörfer 1998, p. 93). Frequently, pro-environmental behavior only reduces environmental harm significantly if others act accordingly. The individual might perceive their own contribution as useless due to a lack of other contributors and therefore has fewer incentives to act themselves. Garrett Hardin's (1968) "tragedy of the commons" opens up another perspective. Environmental goods or environmental quality are (usually) common property. They are open to all, no one can be excluded (Samuelson 1954), for example, from breathing the air. Everyone benefits from clean air but as there is no proprietor no one is really responsible for sustaining it. As the subjective rationality hypothesis is a rational-choice approach the actor has the opportunity to contribute to the commons or to free ride. Rational actors try to maximize their utility. When free riding they benefit from the commons without costs whereas contribution to the commons entails costs. This, in return, lowers the benefits and therefore free riding would be preferred by an instrumental-rational actor as it is hoped or assumed that enough others contribute. The incentive to free ride increases along with an increase in group size (Clark *et al.* 2003). Consequently, incentives to contribute are small (Olson 1965). The subjective rationality strategy thus justifies environmentally unfriendly behavior either by denying its effect or by viewing it as dispensable. The perceived reality (see chapter 3.1.1) is, so to say, construed in a way that it justifies not contributing. Rationality, thus, is subjective and can be changed or adjusted when new knowledge is acquired, when costs occur that had not been in the equation up to that point or when costs are found to be actually lower than expected.

This also entails a contradiction between individual or corporate (i.e. of a company etc.) and collective rationality on the one hand and between long- and short-term rationality on the other hand (Preisendörfer 1996a). Individual and short-term utilities are usually valued higher, and are hence the choices preferred (Hardin 1971) thereby leading to non-contribution. To give a brief example of a corporate actor: the NAIIC (Nuclear Accident Independent Investigation Commission) who investigated the nuclear disaster of Fukushima attested that, at least in part, organizational self-interest was valued higher than the public safety-interest and thus important measures were not put in place to prevent such accidents (Nuclear Accident Independent Investigation Commission (NAIIC) 2012, p. 16).

As environmentally benign behavior often serves the collective interest more than the individual's and since the positive effects generally are settled in the future, people or organizations consequently behave in a less environmentally friendly way even though they might hold pro-environmental attitudes. This is justified by subjective rationality 'subterfuge'. Theorizing on less environmentally friendly conduct via a subjective rationality is not the only way to do so; there are other explanations for not acting in a pro-environmental way despite one's values: the low-cost hypothesis is explained in the following section and the attention-shifting strategy described thereafter.

Low-cost hypothesis

As more generally discussed in the previous section there are factors that inhibit attitudes from turning automatically into respective action. This is also true for environmentally benign attitudes and behavior (Dunlap 1989, Winter 2000, Nisbet *et al.* 2009, Wray-Lake *et al.* 2010) and will be specified as follows. The low-cost hypothesis (e.g. Diekmann and Preisendörfer 1998, 2003, Gatersleben *et al.* 2002) originates from the more general rational-choice theory, but aims at one particular mechanism. It was developed to help explain why attitudes and respective behavior correlate rather weakly and why there is a lot of variation between studies on environmental behavior (Best and Kroneberg 2012).

The basic assumption of the low-cost hypothesis is that as costs of a certain action rise the influence of attitudes on that action declines. The impact of attitudes is moderated by the action's costs. These costs include not only monetary expenses but also spending of time, loss of comfort or "other subjectively defined inhibitions" (Diekmann and Preisendörfer 1998, p. 89). Subjective environmental awareness and concern alone, therefore, do not suffice to explain environmental behavior. Recently, Tobler et al. (2012, p. 204) have shown "that people prefer to show their environmental concern in low-cost areas". Hence, "external conditions" (Guagnano *et al.* 1995, p. 705) determine the influence of pro-environmental attitudes on behavior. Or in statistical terms: attitudes will correlate stronger with low- than with high-cost actions.

Schan (2000) criticizes that the relationship between attitudes and respective behavior moderated by behavioral costs is not (negatively) linear but actually follows an inverted U-shaped path. Accordingly, very "cheap" actions do not need pro-environmental attitudes and thus the relationship is low: people simply act in such a way in their everyday routine or as a "normative function" in which "environmental concern is not a prerequisite for action" (Walton and Austin 2011, p. 274). With rising costs, attitudes play a more and more important part and are translated into action. When actions become too costly, the effect of attitudes on the behavior declines again and extrinsic motivational factors come into play. Either with linear correlation or an inverted U-shape curve, costs seem to influence the effect of attitudes on behavior. Therefore this study will look at different types of behavior.

Attention-shifting strategy

As with the low-cost hypothesis and subjective-rationality considerations, the attention-shifting-hypothesis is another rational-choice explanation for environmental behavior. According to the attention-shifting strategy (Diekmann and

Preisendörfer 1998) the utility of pro-environmental behavior increases and, therefore, is more probable when a reward in the form of positive recognition for the behavior is expected. This especially applies with the phase of enthusiasm of the so-called Issue-Attention Cycle (Downs 1972), when environmental problems are prominent in the public's consciousness and therefore graded higher. This means that actions with the highest 'reputation' are upgraded and conducted, as a means of "ecological correctness" (Diekmann and Preisendörfer 1998, p. 88, 2003, p. 466) while others less valued are downgraded and ignored. Such higher regared actions have symbolic character outwardly expressing an individual's environmental concern and are therefore preferred (Poortinga *et al.* 2003, p. 61). Downgraded actions are argued against via a mixture "of many attention-shifting arguments which include justifications, excuses and subjective saliencies that are creative or convincing to greater or lesser degrees, for the reconciliation of seemingly inconsistent environmental attitudes and behaviours" (Diekmann and Preisendörfer 1998, p. 89).

Following this reasoning, individuals do not choose actions according to their real environmental impact, especially concerning those behaviors that affect their daily life. Instead, they choose intent-oriented behavior, behavior that is perceived as having an impact on the environment (Gatersleben *et al.* 2002, Stern 2000). Intent-oriented behavior usually has a high reputation in the public's eyes and the attention is hence shifted to these kinds of actions rather than to actions that have a real or more positive impact on the environment. Finally, it has to be mentioned that people do not only choose an action for its popularity, they also often misjudge the effect of a certain action or the effectiveness of a certain technology in relation to its real outcome, for instance, the energy that can be saved (Attari *et al.* 2010). Popularity might augment the positive valuation of an action independently of the real impact and thus facilitate the action because its utility for the actor is increased.

To summarize, while behavioral reasons, costs, benefits and effects are constructed subjectively people prefer costs to incur collectively and benefits to for the individual (subjective rationality). These different costs of an action require issue specific attitudes for an actor to act. When the peak of costs is reached, attitudes become less important as the action becomes subjectively regarded as too expensive (low-cost hypothesis). The attention, then, is shifted to socially appreciated actions which lower the cost-benefit ratio by adding benefits, e.g. due to appreciation (attention-shifting strategy).

Before turning to more specific theoretical approaches, a final remark on rational decisions and the construction of reality needs to be made. Cost-benefit analyses, especially in a 'narrow version', would presume perfect knowledge to be absolutely rational (Opp 1999). Here it is assumed that complete knowledge of cost and benefits is not available, maybe not possible at all. So, a costly behavior that produces little benefit may be the result of a rational decision making process where something was misjudged. Costs and benefits can simply be miscalculated. This is the point where the rational-choice and constructivist theory intercept one another. The constructed reality provides insight that is only accurate to varying degrees into the 'objective reality' and thus the rational choices one makes are based on a subjective reality with incomplete information.

3.2 Macro and micro explanations of environmentalism

In the following section special theoretical perspectives of import for this study will be presented. The point of departure is a discussion of major macro approaches followed by the micro explanations that are of actual interest and supplemental to the constructivist and rational choice theory.

3.2.1 Macro theories of environmentalism

There exists a bunch of hypotheses predicting environmental attitudes and behavior from a macro sociological point of view. These theories developed from different perspectives or "stress different dimensions of environmental attitudes" (Diekmann and Franzen 1999). Three major hypotheses are well known in the 'community' of environmental sociologists and are, therefore, briefly described: the post-materialism hypothesis, the prosperity and affluence hypothesis and the globalization hypothesis.

These three macro theories will serve as the background for a general comparison of the drivers of environmentalism across countries.

3.2.1.1 The post-materialism hypothesis

The post-materialism hypothesis (Abramson and Inglehart 1995, Inglehart 1990, 1995, 1997) is based on the idea of a general shift in values. Wealthy people can afford the 'luxury' of pro-environmental attitudes and behavior as for them the quality of life as a whole is more important than "economic and physical security" (Abramson and Inglehart 1995, p. 1). As a result of rising affluence in a society, "their members are less preoccupied with the economic struggle for survival and are free to pursue what Inglehart termed post-materialistic goals" (Franzen and Meyer 2010, p. 220). Values emerge "that emphasize higher-order, aesthetic need" (Pampel and Hunter 2012, p. 422). Besides a greater valued quality of life, political freedom or self-fulfillment, one product of this post-materialistic thinking is the desire for a sound and healthy environment; "The rights and beauty of the habitat…" as Bell (2009, p. 24) puts it. According to the post-materialism hypothesis, it is less an intact ecosystem as an important factor for an overall sustainable life of any living form on earth that is valued. Instead the hypothesis puts an emphasis on the quality of life for the individual that is provided by a good environment. Pro-environmental attitudes, as per the post-materialism hypothesis, are part of a general change in values with increasing wealth, they are "subjective values" (Inglehart 1995, p. 57). "Objective problems" (Inglehart 1995, p. 57) on the other hand can also form pro-environmental attitudes as solving these environmental problems is a direct need of the affected people. Environmental concern should, according to the objective problems consideration, be higher where local environmental problems are more pressing. Thus, as a result of his investigation of the World Values Survey, Inglehart (1995) explains environmental concern in developing nations in contrast to post-materialism in developed societies.

Critics argue that post-materialism in general should be addressed as 'idealism' (Huber 2011, p. 99, Rössel 2006), as there seem to be pretty idealistic values behind it:

> "environmentalism seems to represent for Inglehart the idyllic musing of a gilded society – children's passion for stuffed animals, for example, or Ansel Adams's photographs of nature's beauty. Many environmentalists, however, would no doubt respond that ecological threats are far from being issues of postmaterialism – of Maslovian aesthetic self-actualization. Ecological threats are material threats" (Bell 2012, p. 189).

With this said the major aspects of the macro perspective of post-materialism are covered. The issue of post-materialism will be readdressed in the discussion of the micro perspectives on environmentalism.

3.2.1.2 The prosperity and affluence hypothesis

The prosperity and affluence hypothesis assumes "that the quality of the environment is not only a public good but also a good the demand for which rises with income" (Franzen and Meyer 2010, p. 221). Tracing back to environmental economics (Baumol *et al.* 1979, Field 1994) the general assumption is that with increasing wealth of a nation, environmental concern and pro-environmental attitudes and behavior of people increase as well. Consequently, pro-environmental attitudes towards protecting these goods are supposed to be higher in wealthy countries. Additionally, more affluent individuals have more resources to spend on environmental goods. This, of course, also applies to richer people in poor countries but as the absolute amount of monetary individual resources is higher in developed countries the impact of individual prosperity should be greater in these countries. The sum of many affluent individuals aggregates to a high environmental concern at societal level.

Higher environmental concern should be especially true when matters of giving priority to a certain problem (ranking) rather than simply denoting whether it is important or not (rating): richer nations give higher priority to environmental issues than issues like economic growth or poverty, which poorer nations prioritize as they are usually more severe or acute (Diekmann and Franzen 1999, Knight and Messer 2012).

As the relationship between economic growth and environmental degradation is often supposed to follow an inverted U-shaped curve, the environmental Kuznets curve (e.g. Dinda 2004, 2005, Israel and Levinson 2004, McPherson and Nieswiadomy 2005), in developed countries economic growth should produce a reduction of the environmental impact (e.g. due to cleaner technologies). In developing nations environmental degradation will increase until the turning point is reached,

> "one commonly heard explanation for the observed inverse-U-shaped pollution-income path is that it reflects the natural progression of economic development, from cleaner agrarian economies to dirty industrial economies to clean service economies" (Israel and Levinson 2004, p. 2).

Recent research, though, shows that the ecological intensity of human well-being (EIWB) which goes beyond measuring economic activity (Gross Domestic Product: GDP) is U-shaped and thus runs counter to the environmental Kuznets curve assumption (Dietz *et al.* 2012). A monotonically increasing relation between wealth an environmental impact has also been discovered (Wang *et al.* 2013). Why is this excursion to the EKC important? First of all, various scholars argue that one reason for the reduced environmental impact after a tipping point of wealth is a higher level of concern in wealthy countries (e.g. Dinda 2004, 2005, Kijima *et al.* 2010, Wang *et al.* 2013), however, without quoting evidence. Second, the idea of the environmental Kuznets curve can be transferred to the relationship between a nation's economic development and its environmental concern, i.e. first environmental concern increases with GDP and then after a tipping point it decreases again. A recent study also showed that different dimensions of environmental concern are not consistently related to GDP per capita (Knight and Messer 2012). For instance, willingness to pay for environmental protection and the perception of local environmental degradation was negatively correlated to GDP per capita, whereas perceived global seriousness of environmental problems and the preference of environmental protection over economic growth were positively correlated. For the present study, similarly, non-linear and aspect-of-environmentalism-dependent outcomes are expected.

3.2.1.3 *Globalization hypothesis*

Catton and Dunlap (1978, 1979) and Dunlap (2002) developed the New Ecological Paradigm (NEP) or the New Environmental Paradigm (Dunlap and Van Liere 2008) to illustrate the interdependence of human society and the biophysical environment (see Chapter 1). The authors also explain that this relationship has become more and more obvious to people. As a consequence another hypothesis explaining people's pro-environmental attitudes and behavior is established. The globalization hypothesis (Dunlap and Mertig 1995, 1997, Dunlap and York 2008) challenges the post-materialism hypothesis and the prosperity and affluence hypothesis. Environmental concern has become a global phenomenon and is no longer restricted to wealthy nations. "Postmaterialist values would predict consistently positive relationships between citizen concern for the environment and levels of national affluence, and clearly the three waves of the WVS do not produce supportive evidence for either" (Dunlap and York 2008, p. 550). A paradigm shift (Dunlap and Van Liere 1978) has taken place which led to a prevalence of relatively high levels of pro-environmental attitudes in both, poor and wealthy nations, remaining comparatively stable during the last 20–25 years. "Scholars now generally agree that environmental concern is widespread and long lasting, although variable in intensity over time and from place to place and person to person" (Bell 2012, p. 182). A crucial point of the authors' argumentation is that it is not a single factor such as post-materialism or wealth that creates environmentalism. It is more a combination of these factors (post-materialism or wealth) as well as other factors such as the degree of environmental degradation.

Each of these three hypotheses seems logical and an ample amount of empirical evidence has been found for each of hypothesis. Some of them directly test the respective hypotheses, others do not directly address a certain assumption but the results can be transferred to one of the hypotheses. Wray-Lake et al. (2010) report less pro-environmental attitudes in materialist adolescents. Inglehart himself (1995, 1997) as well as Dekker et al. (1997), Duroy (2005), Franzen and Meyer (2010), Gelissen (2007) and Meyer and Liebe (2010) find evidence for the post-materialism hypothesis. The latter two additionally support the prosperity and affluence hypothesis as do Diekmann and Franzen (1999), Franzen (2003), Franzen and Vogl (2013a, 2013b), Freymeyer and Johnson (2010), Kemmelmeier et al. (2002) or Meyer and Liebe (2010). Support for the globalization hypothesis is found, for example, by Brechin and Kempton (1994), Brechin

(1999), Dunlap and York (2008), Guha (2000), Ignatow (2006a) and Longhofer and Schofer (2010). In a recent study, Pampel and Hunter (2012) conclude that these competing theories are incomplete on their own showing that for different age cohorts[4] contrary effects on environmental concern were found. For some, affluence had a positive effect and for others a negative one. At the same time some degree of diffusion took place in other cohorts, i.e. the support for environmental spending was equally high or even higher in low socio economic status groups than in high status groups.

In many cases, the results of the mentioned studies do not support the hypotheses completely or do so only under certain restrictions. One or two of the three hypotheses might find support while the other one is rejected. If considered carefully the, three hypotheses do not absolutely contradict each other. As mentioned above, they emphasize different dimensions from different perspectives at a very general level. As the results of the studies and the evidence for the hypotheses differ significantly (Franzen and Vogl 2010) the development of an approach integrating different hypotheses is the concluding step to take: "studies using single theories omit crucial explanatory variables and, hence, might be misleading" (Liebe *et al.* 2011, p. 106). Being misled by one-sided or narrow research outcomes may result in policies that do not cover all aspects influencing environmental behavior and thus "often fall short of the objectives" (Clark *et al.* 2003, p. 238).

To summarize, there is no single explanation of environmentalism (Hadler and Haller 2011) and, for that reason, a comprehensive theoretical framework explaining environmental behavior will be extensive rather than simple and reduced to a single issue. To disaggregate a structure into smaller parts will help to understand the structure on a basic level but leaves out "important emergent relationships out of the analysis" (York and Clark 2010). This cannot simply be done by aggregating the knowledge found in the disaggregated structures (Gould 2003, York and Clark 2010). The factors influencing the environmental attitudes and behavior are so manifold that a deeper analysis of the whole process of the formation of pro-environmental attitudes and behavior needs to be conducted. This leads to the next part, the micro theories of environmentalism.

4 The authors classify the age cohorts into 5 year age groups: 25–29, 30–34 etc. (Pampel and Hunter 2012, p. 431)

3.2.2 Micro theories of environmentalism

The focuses in this study are the individual's pro-environmental attitudes and behavior. Important factors that are supposed to explain both are described subsequently. Basic theoretical assumptions and important research results are presented in this part.

After explaining why the globalization hypothesis also has some 'micro relevance', the different approaches will be classified as follows: gender and age, social position and general values. Recalling the constructivist and rational choice approach (3.1), an individual's values as well as social position influence the way in which the environment is perceived, evaluated, and finally acted towards. These values and social position open thought patterns, and offer material as well as mental incentives and constraints.

3.2.2.1 Globalization hypothesis at micro level: virtual environmentalism

Given that environmental concern has become globalized and is thus present in poor, developing and developed nations, it is of interest to see to what extent this omnipresent concern is formed differently in different parts of the world. The globally ubiquitous concern and its variation might have quite different origins in different nations. On a national level, Bell (2012, pp. 283–286) would speak of "virtual environmentalism", meaning to make environmentalism easy so that it is less dependent on values or positions, and with regard to environmental action, less dependent on people's environmental concern. When environment and environmental protection has a broad acceptance in society (social-structural) and at the same time is easily available (infra-structural) it should have 'globalized' or diffused into society. To give an example, this would manifest in a high amount of pro-environmental action within a country and at the same time only small contribution of values, social positions or environmental attitudes to this action.

3.2.2.2 Age and gender effects

Age effects can be explained by different societal as well as environmental conditions that influence socialization and thus environmental attitudes and behavior (Kanagy *et al.* 1994). More generally, a life-cycle effect (Gelissen 2007, p. 99) leads to more idealistic attitudes (such as pro-environmental attitudes) in younger years which decline when people get older. Via the media or direct experience, younger people also have more contact with environmental issues than older people, and, therefore exhibit a higher degree of pro-environmental attitudes (Franzen and

Vogl 2010, p. 222). Research has shown relatively consistent results for age with regard to environmental attitudes (Dietz *et al.* 1998, Gelissen 2007, Liere and Dunlap 1980). Younger people tend to be more environmentally concerned than older people. However, research also unveiled that, despite their lower concern, older people are more often engaged in pro-environmental behavior (Engel and Pötschke 1998, Hadler and Haller 2011, Olofsson and Öhman 2006, Olsson 1994, Preisendörfer 1996a, Swami *et al.* 2011). Blaikie (1992) argues that age effects follow an inverted U-shape meaning that younger people are less environmentally friendly and as people's age rises environmental friendliness also increases until a turning point is reached and it then decreases again.

Women are often found to be more environmentally concerned than men (e.g. Franzen and Meyer 2010, Gelissen 2007, Givens and Jorgenson 2011). Three main explanations attribute to the higher level of environmental concern and respective behavior of women: "women's traditional gender socialization, a motherhood mentality or an ethic of care…" (Gelissen 2007, p. 399). Men in contrast are more instrumental-rational as a consequence of being the traditional "economic provider" (Gelissen 2007, p. 399). The most effective ways to provide high economic outcomes are often quite un-ecological and these negative effects for the environment are ignored (Blocker and Eckberg 1997, Gelissen 2007). Despite the claim that women hold higher environmental concern, the results for gender are still somewhat more ambiguous than those for age. Some find women more environmentally friendly than men (e.g. Franzen and Meyer 2010, Gelissen 2007, Givens and Jorgenson 2011, Olsson 1994), other authors do not find any differences (e.g. Hayes 2001, Tobler *et al.* 2012). In van Liere and Dunlap (1980) women are found to be less environmentally friendly than men. Swamii et al. (2011, p. 25) find these gender effects, too, but they are indirect effects mediated via different personality traits which in turn lead to pro-environmental behavior.

Based on research, the effects of age are supposed to follow an inverted u-shape while for gender no directed hypothesis can be stated. The simple hypothesis would be: men and women differ in terms of their relationship to environmentalism.

3.2.2.3 Social positions

Individual wealth

With reference to the prosperity and affluence hypothesis and the wealth-of-nations-assumptions on the macro-level, an individual's wealth is supposed to have a direct influence on pro-environmental attitudes and behavior. The possible macro outcomes of wealth and affluence, as discussed in the macro section, are strongly dependent on individual wealth. More resources are available to satisfy the higher demand for environmental goods. The assumption is that wealthier individuals are more concerned and more engaged in pro-environmental action. On the other hand individuals also have more demand for other goods or hold status positions in which environmentalism may be socially approved or disapproved of etc.

In the same way that the explanation of why individual wealth should influence environmentalism indicates a positive, negative or zero effect, the state of research can be presented. Non-significant coefficients for income are found in Nawrotzki (2012), and Welte and Anastasio (Welte and Anastasio 2010) do not find an effect of status. A positive influence of an individual's wealth is discovered inter alia in Franzen and Meyer (2010), Freymeyer and Johnson (2010), Givens and Jorgenson (2011), Meyer and Liebe (2010) or Liebe et al. (2011). Concerning environmental political behavior, the effect is also positive in Hadler and Haller (2011). In contrast, the wealth effect in Hadler and Haller (2011) is negative for everyday behavior. Other negative effects of individual wealth on environmentalism are found in Pepper et al. (2011).

Biographical availability assumption

To behave in an environmentally friendly way or inform oneself about environmental issues often requires a certain amount of time. Using public transport instead of a car might turn 15–20 minutes into half an hour whereas going to a demonstration requires several hours of spare time that a person needs to be willing to spend on this activity. This theoretical concept is called the biographical availability assumption (Gillham 2008, Hadler and Haller 2011, Hunter *et al.* 2004). Time-consuming activities that have priority like caring for children, having a full-time job or the responsibilities that come with a marriage (Hadler and Haller 2011, p. 321) can hinder people from acting in an environmentally benign fashion. Contrary to pro-environmental activities, having a job or caring for children are non-voluntary activities. There is no time left for environmentalism, be it objectively having no time left or not being willing to spend the little time that is left on environmentalism. Counterarguments, however, are intuitive. Having children may increase environmental concern or behavior in order to reduce one's environmental impact for the future generation. In accordance with the wealth assumption, being in a full-time occupation would lead to a higher income and thus to a higher level of environmentalism. Also it is logical that a more secure life situation (i.e. having a paid job) opens the opportunity to deal with other issues such as the environment instead of being primarily bothered with one's own instable economic situation.

Research findings are also contradictory. Hadler and Haller (2011) show that marital status indeed has a negative effect on private environmentally-friendly behavior while an individual's economic activity increases this private behavior. Regarding pubic behavior, neither of these measures exerts an influence. The coefficients of children in the household are mostly non-significant in Gillham (2008) with the exception of the final model where a variety of micro and macro predictors are included. This implies that having children leads to several other aspects that both positively and negatively influence environmental activism and thus a zero-effect is encountered when not all relevant covariates are controlled for. In earlier studies, employed people are also more likely to engage in the activities of social movements than unemployed people(McAdam 1986, Nepstad and Smith 1999).

The outcome of the biographic availability assumption is very uncertain. It can be positive, negative or maybe the positive and negative aspects counter each other and so it may yield a zero-effect.

Education and knowledge deficit theory

In general, formal education has been found to be a relatively consistent predictor of environmental concern (Buttel 1987, Ignatow 2006a, Klineberg *et al.* 1998, Meyer and Liebe 2010, Pampel and Hunter 2012); "we find that education is the single, most stable variable that explains environmental concern" (Olofsson and Öhman 2006, p. 784). It is suggested that better education leads to a greater ability to reflect and understand complex issues such as environmental problems and threats (Dunlap and Van Liere 2008, p. 25) as well as to improve perception, awareness and understanding of laws related to environmental issues (Akella and Cannon 2009, p. 533). Therefore a central role in the formation of environmental attitudes and behavior is attributed to education.

Another approach complements the effects of education. The knowledge-deficit theory (Schulz 2002) assumes a certain relation between behavior and the knowledge concerning the topic the behavior is related to. "The primary assumption of knowledge-deficit theory is that people want to help [...] but lack information about how and/or why they should make changes in their behavior" (Nolan 2010, p. 644 f.). Various studies investigated the effects of environmental knowledge on environmental concern and action (Dolnicar and Grün 2009, Franzen and Meyer 2010, Kellstedt *et al.* 2008, Nolan 2010). Unfortunately the ISSP 2010 does not provide a sufficient range of items that measure environmental knowledge. Consequently, the assumption that knowledge leads to more environmental action cannot be tested. Due to its importance, the basic idea of this approach has nevertheless been outlined here.

Population and urbanization

"Cities are social melting pots; sites of innovation, political engagement, and cultural interchange; and drivers of social change. However, cities are also marked by social differentiation, poverty, conflict, and environmental degradation" (Beall *et al.* 2010, p. 3). Based on Jahn (1998) in densely populated regions environmental problems are more obvious than in places with less population density. Environmental bads concentrate on one place and more people at the same time are affected by the resulting problems. Important environmental problems of cities include the quality of air and water, waste removal or stress caused by noise (Vlahov and Galea 2002) as well as habitat loss and threats to biodiversity (Kowarik 2011).

The United Nations predicts further growth of urban populations by 2.6 billion over the next decades (United Nations 2010), and the pace of urbanization is also expected to increase (Vlahov and Galea 2002). With a growing population the pressure increases as more people concentrate more bads on one place and need more resources to satisfy their needs which also impacts areas outside the city's boundaries (Beall *et al.* 2010, Rees 1992). As well as new humans constantly being born, migration to the cities in particular also leads to urban population growth. "Urbanization refers to the change in size, density, and heterogeneity of cities." (Vlahov and Galea 2002, p. 4). Thus, on a global scale, an ongoing process of urbanization piles the pressure on land and nature. Liddle and Lung (2010), for instance, find positive influences of both, the total population (middle aged) and urbanization on different kinds of carbon dioxide exhaust and energy use. Schubert et al. (2013) also discover positive (i.e. a reduction) effects on carbon emission in urban areas due to physical concentration. People living wall to wall and in smaller homes live in a more energy-efficient way. This is unfortunately counteracted by a lower social concentration, i.e. a lower number of household members.

As a place of innovation, growth and change in the size of the community people live in has an impact on their environmentalism. Israel and Levinson (2004), for instance find people living in larger communities to have a higher marginal willingness to pay for environmental protection, Hadler and Haller (2011) discover a positive effect of community size on everyday behavior. Besides these discoveries, the perception of environmental threats is especially seen to increase when living in a densely populated area, in particular in developing countries where objective environmental problems are often more severe (Adeola 2000).

3.2.2.4 General values

Post- materialism hypothesis

The basic ideas of post-materialism have been described in the 'macro section' above. Post-materialism can be a driving force to alleviate pro-environmental attitudes and behavior. It is hypothesized that this effect tends to take place on the micro level than on the macro level (e.g. Davis 2000). Even though Davis (2000) does not find the effects he hypothesized for post-materialism on the micro level, this assumption is supported by the findings, for instance, of Franzen and Meyer (2010)(2010), Kemmelmeier et al. (2002), Kidd and Lee (1997), Meyer and Liebe (2010) or Olofsson and Öhman (2006). Accordingly, an individual's post-materialism plays a more important role than a society's percentage of post-materialists as predictor of environmentalism. Following this postulation, it is logical to suggest that this relationship will hold true in both developed and developing countries.

As already brought up in the macro perspective of post-materialism, Inglehart's (1990, 1995, 1997) idea of environmentalism is very idealistic. Because there is also a materialist aspect to environmental problems, post-materialism should rather be capable of positively influencing the idealist part of environmental concern, in this case the evaluation of environmental issues or environmental activism. The more rational or materialist dimensions, the perception of environmental threats, willingness to make sacrifices, everyday action or the acceptance of laws should not be influenced or at least less strongly.

Political position

Certain values are closely related to certain political positions or views a person holds. Concerning environmental attitudes the green parties in particular mirror this relationship. As the green parties are usually on the left and more liberal side of the political spectrum, more liberal people on the left wing have generally been found to be more environmentally concerned (e.g. Barr 2007, Engel and Pötschke 1998, Jones and Dunlap 1992, Olofsson and Öhman 2006, Tobler *et al.* 2012, Witzke and Urfei 2001). Nawortzki (2012), in contrast, discovered a varying influence of conservative world views on environmental concern as "a function of country-level characteristics" (p. 286): in developed societies, conservatives are less concerned than liberals. In former communist nations, conservatives are more concerned, as is the case in countries with worse environmental conditions (p. 301).

For the present investigation it is assumed that the further to the left people's world view is, the greater their level of environmentalism. In countries with a communist history, conservatives are more likely to be more environmentally friendly.

Generalized trust and conditional cooperation

Complementing formal norms, "trust works as a foundation for society upon which planning, justice etc. are only possible" (Heinrichs and Gross 2010, p. 10). An individual's behavior is not always independent of the behavior of others. The theory of generalized trust (Meyer and Liebe 2010, Ostrom 2000) assumes that people with a higher level of trust in others are more willing to do something themselves. Trust motivates one's cooperation (Walter 2010, p. 129) but it is relatively independent of the actual behavior of others. By contrast, conditional cooperators (Fischbacher *et al.* 2001, Gächter 2007) only contribute to a public good, "when they are convinced, that others are also doing so" (Meyer and Liebe 2010, p. 47). Therefore, people who show a high level of trust in others behave in a more environmentally friendly way than those who look askance at others or do not behave independently of others. However, it is not only the level of action where trust is expected to reveal its influence: "more generally, the decline in trust reflects a growing pessimism among Americans" (Uslaner 2002, p. 7). This quote entails that more trust goes hand in hand with a higher degree of optimism. Consequently, more trustful people should be more optimistic and likewise also less skeptical about environmental issues. This, in turn, should increase environmental concern.

As trust is fundamental to democratic societies (Jalava 2003, Lewis and Weigert 1985, Misztal 2001), trust in society's institutions, e.g. political system, the economy, the legal system etc., is crucial, too (Walter 2010). Yet, the people's trust in institutional decisions, policies, legal regulation, etc. could hamper the individual's behavior. A 'they will handle it' tendency might arise with too much trust. The impression that personal action is not necessary is then the consequence. Hence, to put all the trust in policy and other decision makers and thus transfer the responsibility into their hands can lead to reduced environmental behavior.

Generally, for the present study, trust is expected to increase environmentalism.

Because different countries from various cultural, economic and ecologic regions and backgrounds are investigated here, differences in strength and direction of the effects are expected in the empirical analysis. For instance, it is quite likely that gender affects environmental concern differently depending on the country under investigation. Statistically one would say that random intercept and random slope are expected.

3.3 Environmentalism

After having spent so much words on environmental problems, the reason why studying environmental attitudes and behavior as well as theorizing the factors that actually influence environmentalism, it is time to explain what in particular can be understood when the terms environmental attitudes, environmental behavior, action or activism are mentioned.

3.3.1 Environmental perception, evaluation and willingness to pay

According to Ignatow (2006a) there are two general environmental worldviews. The ecological worldview which emphasizes the interdependency of humans and nature forms a secular, rather scientific perspective. On the other side there is the spiritual model of environmentalism where nature is sacred, made by God, and is threatened by science and technology (Ignatow 2006a, p. 443). In this study only the ecological, secular perspective of environmentalism will be investigated.

In the literature dealing with these attitudes, three dimensions of environmental concern have been elaborated: the *cognitive*, the *affective* and the *conative* (intentional) dimension or component (Franzen and Meyer 2010, Maloney and Ward 1973, Maloney *et al.* 1975). The *cognitive* component covers the rational perception of and knowledge about environmental problems. The *affective* dimension covers the emotional reaction toward them. Huber (2011, p. 81) summarizes the three components and combines an *evaluative* element with the affective dimension. For the present study the *evaluative* aspect is better covered by the questions in the ISSP than the affective one (see chapter 4.2.3), therefore, it is referred to as the *evaluative* dimension. As the name indicates, in this dimension, environmental issues are evaluated. This includes environmentally skeptical views, evaluation of the importance of environmental protection or the individual's opinion whether something can or should be done. The *conative* component covers the intention to do something. This component usually is referred to as marginal or in-principle willingness to pay for environmental protection (Liebe *et al.* 2011, Torgler and García-Valiñas 2007).

Various articles investigate environmental concern and how it is formed (Barr 2007, Dunlap and York 2008, Franzen and Meyer 2010, Gelissen 2007, Ignatow 2006a, Olofsson and Öhman 2006). Its major predictors were discussed in the previous chapter. Pro-environmental attitudes are regularly found to positively influence pro-environmental behavior but as for example postulated by the low-cost hypothesis or attention shifting strategy the size of the effects varies (e.g. Bamberg and Möser 2007, Best 2009, Best and Kneip 2011, Diekmann and Preisendörfer 1998, 2003, Gatersleben *et al.* 2002, Tobler *et al.* 2012).

3.3.2 Environmental action

Environmentally significant behavior can be defined in the words of Stern (2000, p. 408): "the extent to which it changes the availability of materials or energy from the environment or alters the structure and dynamics of ecosystems or the biosphere itself." This broad impact-oriented definition is complemented by an intent-oriented definition directed more at pro-environmental behavior: "behaviour that is undertaken with the intention to change (normally, to benefit) the environment" (Stern 2000, p. 408). Hadler and Haller (2011) differentiate between public and private pro-environmental behavior in their analysis of the ISSP 2000. The authors refer to Stern (2000) who separates public behavior into activism, "extremely active participation in social movements" (Hadler and Haller 2011, p. 317) and less extreme forms of public behavior such as signing a petition. In their study as well as in Hunter et al. (2004) and in the present study (see Chapter 4.2.3) the separation of public behavior cannot be validated empirically and therefore these two types are combined into one aspect: environmental activism. Environmental activism only has an indirect impact on the environment as it is meant to urge other actors (e.g. policy-makers) to do something. Stern (2000, p. 409) also mentions support for public policies as a form of non-activist behavior. As the analysis will show in the ISSP the support for laws is a dimension separated from activism.

Laws and state regulations may play an important role in forming an individual's attitude behavior. They have similarly important implications for the environment-society relation as a whole. This accompanies the problems concerning implementation and enforcement discussed in chapter 2.2. Laws that create obligatory actions or prohibit other activities are good examples. For instance, if driving a car on weekends was made illegal to reduce pollution through car exhaust, certainly most people would cut back on driving a car on weekends (or they would go out on the streets and demonstrate against such a law). They would act in a more environmentally friendly way as a consequence of formalization of behavior. Formal rules (like laws) are crucial for a democratic society to work (Misztal 2001). They are also an important instrument to improve environmental protection. However, "the process of formalization, by withdrawing responsibility for negotiation from participants, may reduce their mutual trust and understanding" (Misztal 2001, p. 337). Responsibility is then seen as belonging to the government and institutions instead of to the individual. This may hamper pro-environmental behavior in other areas, especially in the private sector, as a result of a reduction to be "solely the function of legal and abstract dicta" (Seligman 1997, p. 173). Even though this is a relatively empty spot in the field of research, at least one article includes attitudes toward laws as predictors of environmental behavior. Hadler and Haller (2011) find that being in favor of state interventions in general, not restricted to environmental regulation, also increases public as well as private pro-environmental behavior. For the present study, support for environmental regulation is seen as a very low-cost form of environmental action. Actually, it is non-acting. Supporting laws on the one hand means having the government to deal with environmental problems. On the other hand, this can also imply having the government decide what the right thing for people to do is. So, either people do not have to act at all or at least they do not have to think about which behavior among many alternatives is the right one, and how to do it.

In contrast to activism and the acceptance of laws, private behavior has a direct effect, often environmentally significant "only in the aggregate, when many people independently do the same thing" (Stern 2000, p. 410). In the framework of the present study, private behavior is summarized as the activities of people in their day-to-day routine; hence the term everyday action is preferred. Generally, these activities include consumer decisions, saving (e.g. energy) activities, transportation, waste separation and recycling.

Following the previous discussion, for this study environmental action is separated into three elements: everyday action, activism and the support for environmental laws. So far, the words behavior and action have been used more or less synonymously. The question is whether it is environmental behavior or environmental action (in a Weberian sense, Weber 1978) that is dealt with. Generally it can be both. An activity, environmentally friendly or unfriendly, can be considered behavior (at least in environmental respects), when there is no subjective meaning to the environmental part. One example, for instance having fun driving a car, can be an action as it has a subjective meaning to the person driving around; but at the same time it is 'only' behavior in other dimensions because when seen as environmentally relevant, driving a car has no subjective (environmental) meaning for this person, and hence can be considered environmental behavior. However, when the person starts contemplating the 'environmentality' of his or her activity and then cuts back on car driving, this is environmental action. The same could be said if someone uses very little water (for washing, cooking, etc.), without thinking about it, and simply because no more is needed; this again would be referred to as behavior. Using little water consciously in order to save what might be a scarce resource, in contrary, would be considered as action.

This study will deal with environmentally friendly action. All items of the ISSP aim at something that is done for environmental reasons and, hence, bears a subjective meaning.

3.4 The hypothesized model

Figure 1: Conceptualization of environmental behavior according to Fietkau and Kessel (1981)

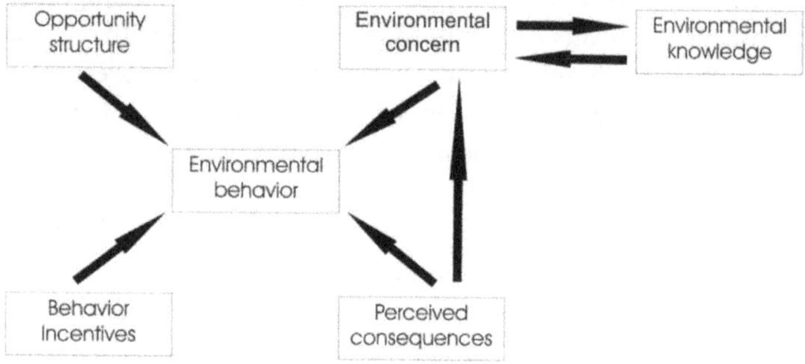

To illustrate how environmental action is related to environmental attitudes and the aspects presented in chapter 3.2, reference can be made to a model by Fietkau and Kessel (1981). The authors conceptualize environmental behavior as presented in Figure 1, but many aspects of the model cannot be fully covered in the present study. As already mentioned in chapter 3.2.2.3, environmental knowledge is not captured appropriately in the ISSP 2010. For perceived consequences, no measures are available either. Integration of opportunity structure and behavioral incentives can also only be indirectly derived from the micro theories. On the one hand, a higher income opens up the opportunity to spend more on environmentally benign goods because they are more expensive in many instances. On the other hand, in many countries, as chapter 4.2.3 will show, behaviors like recycling are infra-structurally not available. Value-rational or instrumentally-rational incentives as discussed in chapter 3.1.2 provide the behavioral incentives.

Figure 2: Path diagram: Environmental attitudes and action

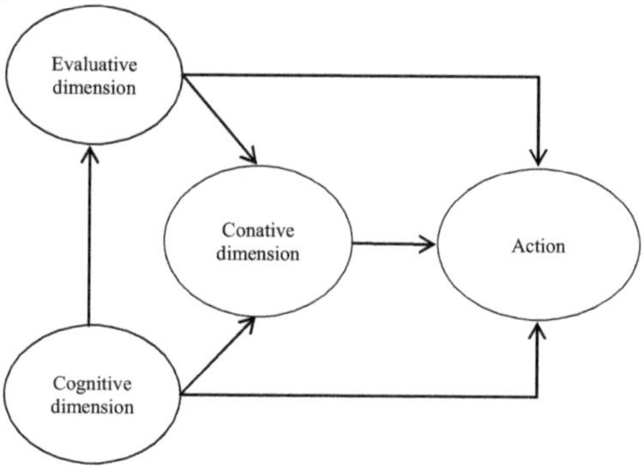

Figures 2 and 3 show the conceptual approach for the present study. These are less abstract images more on the 'variable level', hence they are the graphical representations for the multivariate analysis. Figure 2 illustrates the supposed relationship between the three dimensions of environmental concern and how these are expected to be related to action. The cognition of environmental problems is supposed to affect not only the evaluation and action intention but also the action directly. It is easy to imagine that a person who perceives an environmental threat reacts emotionally/normatively to it. This reaction, from a value-rational perspective, may turn into action intention or action itself because it feels like the right thing to do. From a more instrumental rational perspective the cognition might follow a direct route to the behavioural intention or the action itself. When an environmental problem poses a threat, for instance, to a person's health it is relatively instrumentally rational when trying to prevent this threat that the person attempts to reduce their own environmental impact; and thus contribute to reducing the environmental problem and to protecting one's own health. Whether these assumptions can be supported or not, and whether there are country-specific differences, will be shown in the empirical analysis.

Figure 3: Path diagram - Predictors influencing environmentalism

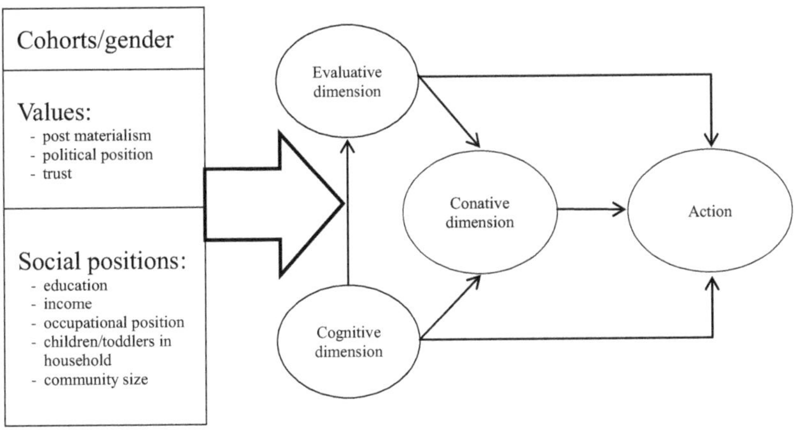

After these hypothesized paths between environmental attitudes and action are tested, further predictors of environmentalism derived from chapter 3 are added to the model as schematically illustrated in Figure 3. From the theoretical background it is not clear if these predictors tie in more with the constructivist perspective by impacting the attitudinal level or if they directly open up opportunities and set constraints or incentives for behavioral decisions; or both. This unclear relationship is represented by the large arrow in Figure 3 pointing at the basic model. Before the substantive empirical analysis can start, the data set used will be described in the next chapter. In this instance the issue of data quality and the operationalization of the dependent and independent variables will be addressed as well.

4. Data, data quality and measures

For the empirical analysis, the data of the International Social Survey Programme (ISSP) Environment III will be utilized. After the years 1993 and 2000, the environment was the subject of the ISSP for the third time in 2010. The 'ISSP community' has now grown to 47 participating countries in 2013. Of these 47 countries, 32 participated in the 2010 module. Due to the separated samples of Eastern and Western Germany, as well as Israel (Jews) and Israel (Arabs), 34 samples are available. The subjects of this annual survey range from national identity, to sport and leisure, the role of the government and the environment, with each topic reoccurring every 5–10 years in order to trace developments over time. In Table A1 in the appendix, all participating countries are listed in alphabetical order. In addition to socio-demographic variables, the ISSP provides a broad range of items related to different dimensions of attitudes and behavior concerning the environment. The data can be downloaded via Zacat, the Online Study Catalogue of GESIS – Leibnitz Institute for the Social Sciences.[5] For further information about the ISSP in general, consult www.issp.org.

The World Bank (www.worldbank.org) provides the data on GDP per capita (current US$) required for the macro analysis and the International Energy Agency (2012) offers country-specific data on total primary energy supply (TPES) which is used in the country profiles to compare people's preferences with the actual supply situation.

Instead of pursuing a multilevel approach, the macro and micro analysis will proceed in two separate steps. Although the focus is primarily on the micro determinants of environmentalism in the framework of a cross-national comparative study, it is reasonable to have a look at differences on the macro level as well. Therefore, it was decided to conduct country-specific analysis (multi-group analysis) on the micro level preceded by a broad macro assessment. The advantage over a multilevel analysis is that country-specific results are obtained. Instead of having one regression coefficient, say from equivalence income on environmental action with a variance that is significant (as in a random slope model), a coefficient (or slope) is estimated for each country. The researcher obtains country-specific numbers for each characteristic (such as R^2) that might be of interest. Thus, a more detailed look into country features is provided. Additionally, multi-group analysis allows testing for significant differences of the coefficients between groups, i.e. does the

5 www.gesis.org; zacat.gesis.org/webview/index.jsp.

effect of trust on environmental concern differ significantly between Eastern and Western Germany?[6] Unquestionably, there are disadvantages with this approach. In a multilevel analysis the results can be easily summarized in one table. With the multi-group method the researcher needs one table for each country and consequently also country-specific interpretations of the results. In the present study six nations will be compared. This results in six descriptions and interpretations, for instance of the effects of education on the outcome variables. Since the analytical advantages outweigh the formal disadvantages and as this method is more appropriate for the present aim of the study, the country-specific multi-group analysis will be the method of choice.

4.1 Methods

To assess the quality of the data, the frequencies of missing values will be put under the microscope, first. For an investigation of the quality and comparability of the scales that are supposed to measure environmental concern and action Multiple Correspondence Analysis (MCA) will be applied (Blasius and Thiessen 2012). The MCA is an analysis of contingency tables and visualizes the item's categories in a multi-dimensional space. The MCA also yields factor loadings and eigenvalues but in the present application the graphical output is of interest. The substantive variation in the data should be mirrored on the abscissa (1st dimension) while methodologically induced variation is reflected on the ordinate (2nd dimension). On the first dimension the item's categories should retain their order and on the second dimension categories should form a horseshoe with the more extreme categories further in the positive or negative area of the second dimension. Graphical examples will be provided in the application of the MCA. This method allows the user to see if the order of categories is indeed retained or if there are changes, indicating that the ordinality of the items within the scale is violated. This in turn speaks for inconsistencies in the responses. If the ordinality is retained on the second dimension while the first dimension mirrors the horseshoe, the methodologically induced variation is higher than the substantive variation (Blasius and Greenacre 2006, p. 9). This is an indication of poor data quality (Blasius and Thiessen 2006, p. 238).

After the assessment of data quality for each dimension of environmental concern and action a categorical principal component analysis (CatPCA) is conduct-

6 It is also possible to test for significance of coefficients within one group, for instance if the coefficient of those from the center-left political spectrum differs significantly from the value of the far-left.

ed, one for each country where the data is of good quality. In CatPCA (also called non-linear principal component analysis) the original categories of the items are replaced by optimal scores (also called quantifications) in an iterative process (Blasius and Gower 2005, p. 374). This means that the true distances between the categories are approximated, thus transferring the original ordinal scale into an interval scale. Each item can then be saved as a new variable with the rescaled values. The new items have a mean of zero and a standard deviation of one.

As the next step country specific confirmatory factor analysis (CFA) allows testing the hypothesized variable structure of the measures of environmental concern and action. For the CFA the variables rescaled via CatPCA are used. While in the MCA framework only one scale at a time is assessed, a full structural equation model (SEM) will serve to investigate the relations between environmental concern and action as well as the influence of age, gender, general value and social position on these aspects. The advantage of structural equation modeling is that multiple relations that often exist between the independent variables can be modeled easily, as well as mediation effects. Another big advantage over standard regression techniques is that SEM offers the possibility to include several dependent variables in one model. The major dependent variable is environmental action and the independent ones are, for instance, environmental attitudes. These in turn are dependent variables that are regressed on general values. Yet, "values do not act alone" (Dietz *et al.* 2005, p. 356) and therefore social positions and their influence on environmental attitudes and environmental action can be included. This can be done in one model estimating the effects simultaneously.

To test if the model fits the data well, three fit indices are provided. From the high number of fit indices AMOS provides the most widely spread and used were chosen (Iacobucci 2010): The Comparative Fit Index (CFI), the Root Mean Square Error of Approximation (RMSEA) and the Standardized Root Mean Squared Residual (SRMR). Iacobucci (2010, p. 90) states that along with the CFI and SRMR the χ^2 should be reported, the first two being descriptive indices while the latter is an inferential statistic (Iacobucci 2010, p. 91). A large sample size increases the χ^2 statistic and it is often significant which would indicate a bad fit. In fact, in all analyses below, χ^2 is significant. To overcome the problems of χ^2 in large samples, only the following fit indices will be applied.

The CFI compares the hypothesized model with the null-model (when all covariances between the observed variables are zero) and "attempts to adjust for model complexity or parsimony. It does so by including the degrees of freedom" (Iacobucci 2010, p. 91). As a goodness-of-fit index a higher CFI value indicates a better fit with a range from 0–1. Hu and Bentler (1999, p. 27) suggested a value of close to or higher than 0.95 as a good fit.

The SRMR is a badness-of-fit index; lower values indicate a better fit. To calculate the SRMR the square root of the average of the difference between the covariance in the data and the estimated covariance is extracted. The result is an index with a range from 0–1 (Iacobucci 2010, p. 91). Hu and Bentler (1999, p. 27) tested the performance of the SRMR and suggested a general value of SRMR ≤ 0.08 in indicate a good fit.

Finally, the RMSEA, also a badness-of-fit index, compares the covariances in the data and the predicted covariances (Byrne 2009, p. 80) with a range from 0–1. In comparison to the SRMR it includes the number of degrees of freedom and is thus more sensitive to model complexity. A good model fit is achieved with a RMSEA ≤ 0.06 (Hu and Bentler 1999, p. 27). Byrne (Byrne 2009, p. 80) discusses values of ≤ 0.08 as an acceptable fit and RMSEA ≤ 0.05 comprises a good fit.

The different scope and computation of the fit indices and the suggested cut-off criteria will help to ensure that there is no overly large models misspecification so that substantive interpretations and conclusions can be drawn from the models.

In a SEM, variables of interest, e.g. environmental concern, are represented as latent variables with a mean of zero and a standard deviation of one. Usually the latent variable is scaled as the indicator manifest variables (e.g. ranging from 1 to 5 in typically likert-type scales). Due to the fact that the original variables are rescaled in the CatPCA into a metric scale with a mean of zero and a standard deviation of one, the latent variables are also scaled in that way.

As country specific as well as cross country comparable effects are relevant, unstandardized coefficients are presented. Additionally many of the independent variables are dummy variables for which an interpretation of a standard deviation change is not feasible as they can only be zero or one; hence standardized coefficients are not useful for the present purposes. Positive unstandardized coefficients (on any dimension of concern or everyday action) would indicate an increase in relation to the average while a negative coefficient decreases the value of the dependent variable below average. In case of the acceptance of laws and activism it is the other way round: a positive coefficient indicates a decreasing support for laws or a decline in the amount of activism. This is a result of the MCA in which the latent variable for both is created and the items categories are quantified in such a way that a higher value indicates less acceptance of laws or activism (see chapter 4.2.4). Owning to the focus on environmental action, in addition to the direct effects (the unstandardized coefficients in the tables), total effects of the independent variables on environmental action will be presented as well. These are complemented by bootstrapped 95% confidence intervals. To achieve the confidence intervals 2000 samples per country and analysis step were drawn from the total sample to deal with the non-normality of the data (Byrne

2009 chapter 12, Hancock and Liu 2012). Furthermore, Little et al. (Little *et al.* 2007, pp. 362–364) consider bootstrapping one of "the most accurate and powerful approaches to assess indirect effects" and thus to test the total effects' (which are calculated as the product of the direct and indirect effects) significance.

> "The idea behind bootstrapping is that you consider your sample to be your population. From this 'population' you take, say, 500 samples *with replacement*. From each sample you then calculate the required sample statistic. This gives you an empirical sample distribution, and averaging the various sample statistics, you get an estimate of the parameter and an empirical standard error of the estimate" (Blunch 2012, p. 236 italics in original).

All coefficients, fit indices, and explained variances (R^2) are based on maximum likelihood estimation. As suggested by Iacobucci (2010, p. 95) this method performs best and is relatively robust to non-normality.

Before the country specific multivariate analysis will be discussed in the respective profile, the fit indices (CFI, RMSEA and SRMR) of the measurement model (CFA) will be provided. In case they indicate any model misspecification, this will be solved before substantive analysis. If the fit indices indicate a good fit, only the indices will be presented without further discussion. The multivariate analysis of each country proceeds as follows: first, a model investigating gender and cohort effects on environmental attitudes an everyday action will be estimated and presented. Then, four models will be estimated in which the measures of general values (i.e. post-materialism, political position and trust) will be included. In model 1 only post-materialism is added to the basic model (see Figure 2). Model 2 only includes the dummies for the political position and model 3 the index of trust; model 4 is the full model integrating all variables that represent general values. For reasons of space and because there are no substantial differences between the models that include a single aspect and the one with all three included, only the final model is then presented in the table.

Analyzing the effects of the social position six models will be estimated, again, one factor at a time. First income is included, followed by education, occupation, school children/toddlers in the household, and finally community size. Model six includes all five predictors. For the same reasons as for general values, only the model with all social positions is reported. As the theoretical assumption of the wealth hypothesis indicated, there is not necessarily a linear relationship between wealth and environmentalism. For that reason, the squared equivalence income is introduced into the equation in a further step. A significant negative coefficient would indicate a concave trajectory while a positive effect would give a convex curve.

The three first steps are taken to investigate age/gender, values, and social positions separately and thus ignoring the relations they might have to one another. Finally, a full model including all important factors will be estimated. Starting with the model including all values, one social position after another will be added to the model. The last two variables added are those of age and gender which are controlled for in the final model. To save space, only the final model with all variables in the equation is presented. Substantial drops or increases in size of coefficients or significance will be explained in the text and traced back to the predictor that caused this parameter change by relying on the step by step analysis.

This approach may lead to different models in different countries as it is supposed that the relations between the various forms of environmentalism and its predictors differ between cultural and social backgrounds. For model parsimony purposes (see Byrne 2010: 183) after all variables on interest are included, the non-significant ones are removed, one at a time. Thus the tables depict only paths that significantly contribute to the explanation of environmental attitudes or action in each country.

The same steps as just described for everyday action are then repeated to explore people's support for environmental laws and activism. To avoid repeating what was already presented in the analysis of everyday action, the tables only present those paths pointing at the support for laws/activism because those influencing attitudes do not change.[7]

However, before these steps can be taken the data quality needs assessment first. In the following chapter, the reader will also learn how the measures of environmentalism and the independent variables are operationalized.

4.2 Data quality and operationalization

When Blasius and Thiessen (2012, p. 1) state that data quality manifests in an absence of artificial variation they refer to various issues that can compromise data quality: study architecture (e.g. survey design, mode of data collection, number and order of questions, format, etc.), institutional practices of data-collecting agencies (e.g. interviewee training, sampling design, monitoring, etc.) and of course the respondents' behaviors (e.g. verbal skills, retrieve information, fatigue, etc.). Any of these issues can create artificial variation that has no substantive meaning. An assessment of institutional practices, the respondents' characteristics by analyzing

7 Minor differences in the estimates are likely to occur due to the slightly different covariance matrix.

their response behavior on the scales that measure environmental concern as well as an investigation of missing values are part of this process. Lachmann (forthcoming, forthcomingb) applied the techniques suggested by Blasius and Thiessen (2012) with an emphasis on faked or duplicated interviews (institutional practices) and assessed the quality of the scale measuring the conative dimension of environmental concern (see chapter 4.2.3 for the operationalization). On this basis only the issue of missing values and the functioning of the scales of environmental concern not evaluated by Lachmann will follow in the next section.

4.2.1 Missing values

Missing values can occur for various reasons. The respondent might have no opinion on a particular issue and so he or she chooses not to provide an answer or states 'no opinion'. The question asked might not be understood by the respondent and hence no answer is given, or the questionnaire asks for something the interviewed is not willing to give an answer to, for instance, someone might not want to tell the interviewer which political party he or she voted.

The first question addressed concerning missing values is whether there are many missing values at all. Are there countries that have substantially higher missing ratios than others and are there some items on which respondents more often choose not to answer or to answer 'no opinion' (NO)? Table 4.1 shows that the percentage of missing answers on 14 selected items on environmental concern is in total 3.5%. The lowest number of missing values is found in Taiwan (0.9%) and the most missing values are encountered in Denmark (6.0%). The amount of missing values is not very high and there is no one country with an extraordinarily high amount of missing values. In the next step single items will be investigated per country to see if there are items on which missing values gather and if this 'gathering' is country specific.

The marginal distribution of one example item is given in Table 4.2. A relatively high number of missing values within this item can be found in Russia (8.7%) and Japan (7.6%). The other countries lie within the range in Table 4.1.

Table 4.1: Missing ratios

Country	N	Number of responses	Number of missing responses	Percentage of missing answers
Argentina	1130	15820	405	0.026
Austria	1019	14266	350	0.025
Bulgaria	1003	14042	554	0.039
Canada	985	13790	276	0.020
Chile	1436	20104	678	0.034
Croatia	1210	16940	301	0.018
Czech Republic	1428	19992	431	0.022
Denmark	1305	18270	1100	0.060
Finland	1211	16954	627	0.037
Flanders	1142	15988	618	0.039
France	2253	31542	1183	0.038
Germany-East	418	5852	275	0.047
Germany-West	989	13846	694	0.050
Great Britain	928	12992	664	0.051
Israel-Arabs	193	2702	52	0.019
Israel-Jews	1023	14322	559	0.039
Japan	1307	18298	886	0.048
Latvia	1000	14000	623	0.045
Lithuania	1023	14322	676	0.047
Mexico	1637	22918	918	0.040
New Zealand	1172	16408	515	0.031
Norway	1382	19348	809	0.042
Philippines	1200	16800	93	0.006
Russia	1619	22666	1238	0.055
Slovak Republic	1159	16226	595	0.037
Slovenia	1082	15148	394	0.026
South Africa	3112	43568	2174	0.050
South Korea	1576	22064	297	0.013
Spain	2560	35840	1240	0.035
Sweden	1181	16534	632	0.038
Switzerland	1212	16968	208	0.012
Taiwan	2209	30926	268	0.009
Turkey	1665	23310	1128	0.048
United States	1430	20020	775	0.039
Total	45199	632786	22236	0.035

Notes: Numbers are based on 14 items measuring environmental concern: Q10a, Q10c, Q12a, Q12b, Q12c, Q13a, Q13c, Q13d, Q13e, Q14a, Q14b, Q14c, Q14d, Q14e; Source: ISSP 2010, own calculations

There is one item that is of special interest. "Do you think that modifying the genes of certain crops is…?" (Q14f) contains an above-average number of missing values. The items' marginal distribution is shown in Table 4.3. In 11 nations the percentage is over 10%. In Belgium (Flanders) and the USA it reaches 17.9% and 16.9%, respectively, on the 'can't choose' option. In Israel (Jews and Arabs) the refusal to answer was relatively high (15.9% and 14.5%, respectively). The high amount of missing values in this particular item could be explained by the fact that it might be indeed hard to tell whether genetically manipulated crops are harmful or not. The effects of manipulated crops is certainly less ostensible than for example the effect of air pollution. Due to the high amount of missing values it seems reasonable to exclude this item from the analysis. The same tendency also holds for item Q14g 'Nuclear power stations are…' which is also not used in subsequent analyses.

In general, missing values are not a problem of the data set or particular nations. When choosing items for analysis, one has to keep an eye on the missing values as in the example above. Besides that, missing values are no clear indicator of data quality. Blasius and Thiessen (2006) find that missing values are not related to data quality, Schäfer et al. (2005) or Bredl et al. (2012) even report that the amount of missing values is lower in faked than in properly conducted interviews. How missing values will be dealt with is explained in the following section.

Table 4.2: Q10a: We worry too much about the future of the environment and not enough about prices and jobs today (%)

	1 sa	2 a	3 nn	4 d	5 sd	8 can't choose	9 NA
Argentina	4.3	26.4	15.7	45.5	5.4	2.7	0.0
Austria	8.3	22.4	22.0	32.6	13.6	1.1	0.0
Flanders	8.0	28.6	27.1	28.8	4.7	1.2	1.6
Bulgaria	9.3	31.5	18.2	28.0	9.3	3.7	0.0
Canada	6.5	21.1	18.8	38.5	14.2	0.4	0.5
Chile	2.3	29.0	30.2	28.6	4.5	4.8	0.6
Taiwan	3.9	34.8	8.7	47.7	3.5	1.4	0.0
Croatia	10.9	28.3	27.6	25.0	6.4	1.7	0.1
Czech Republic	12.0	26.9	26.0	24.8	8.9	1.4	0.0
Denmark	5.9	21.6	16.0	29.0	22.8	2.3	2.5
Finland	9.8	32.0	17.6	29.4	6.3	3.0	1.9
Germany-West	9.1	17.0	18.6	34.7	15.5	2.3	2.8
Germany-East	9.3	28.5	19.6	28.2	10.8	1.2	2.4
Israel-Jews	10.3	28.9	23.1	27.6	4.8	3.3	2.1
Israel-Arabs	7.8	23.3	15.0	40.9	11.4	1.0	0.5
Japan	5.0	13.0	42.2	21.4	10.7	7.0	0.6
South Korea	4.8	16.5	13.7	39.3	24.2	1.5	0.0
Latvia	17.8	30.4	17.8	23.3	5.4	5.3	0.0
New Zealand	7.8	31.0	19.1	32.3	7.1	0.9	1.7
Norway	4.1	23.4	21.1	39.0	9.4	1.0	2.1
Philippines	17.3	40.4	22.8	14.9	4.3	0.3	0.1
Russia	10.1	24.7	26.9	21.5	8.2	8.7	0.0
Slovak Republic	11.0	21.1	31.1	22.3	10.1	4.0	0.3
Slovenia	9.4	35.4	23.8	25.0	3.9	2.1	0.4
Spain	10.4	31.6	19.1	28.1	7.3	2.8	0.7
Sweden	3.6	17.7	28.0	34.7	12.3	2.9	0.8
Switzerland	3.5	22.7	23.1	41.0	8.3	1.0	0.5
Great Britain	9.1	34.5	21.0	27.8	4.3	1.4	1.9
United States	10.6	31.3	15.7	32.1	8.0	2.1	0.3
Total	8.2	27.0	21.6	30.9	9.0	2.6	0.8

Notes: sa = strongely agree; a = agree; nn = neither agree nor disagree; d = disagree; sd = strongly disagree; Source: ISSP 2010, own calculations

Table 4.3: Q14f: Do you think that modifying the genes of certain crops is...

	1 sa	2 a	3 nn	4 d	5 sd	8 can't choose	9 NA
Argentina	17.1	41.7	23.5	6.4	2.0	9.3	0.0
Austria	30.0	34.7	18.2	8.5	3.1	5.4	0.0
Flanders	5.9	14.9	35.0	23.1	3.0	17.9	0.3
Bulgaria	28.2	34.1	19.5	5.6	1.4	11.2	0.0
Canada	13.7	22.1	31.6	19.1	4.5	8.4	0.6
Chile	30.8	42.1	15.7	3.1	0.9	7.3	0.1
Taiwan	6.2	26.9	40.4	14.5	2.5	9.5	0.0
Croatia	25.4	37.4	24.5	7.6	1.6	3.5	0.1
Czech Republic	13.8	25.4	30.0	13.9	4.8	12.3	0.0
Denmark	9.3	17.8	25.7	24.0	5.3	14.8	3.2
Finland	12.2	24.9	24.4	22.5	4.5	9.9	1.6
Germany-West	23.5	35.6	19.6	9.2	1.3	8.9	1.9
Germany-East	22.0	31.1	18.7	14.8	2.4	7.7	3.3
Israel-Jews	10.0	26.1	23.2	17.0	5.2	2.6	15.9
Israel-Arabs	20.2	24.9	21.2	17.1	2.1	0.0	14.5
Japan	12.9	24.3	35.1	11.7	2.4	12.5	1.1
South Korea	17.6	35.5	31.6	10.5	1.0	3.7	0.0
Latvia	19.1	30.5	25.5	11.7	2.0	11.2	0.0
New Zealand	17.0	19.8	29.3	21.2	4.6	6.9	1.2
Norway	6.5	20.5	31.5	23.7	3.7	12.7	1.4
Philippines	16.9	29.3	32.8	13.3	4.3	3.5	0.0
Russia	39.8	30.5	17.7	2.7	0.6	8.7	0.0
Slovak Republic	16.5	31.1	27.4	10.1	2.8	11.3	0.8
Slovenia	20.0	38.4	28.0	6.4	0.9	6.4	0.0
Spain	13.7	32.7	27.6	8.5	2.8	14.2	0.5
Sweden	12.8	22.3	30.2	20.1	2.3	11.8	0.6
Switzerland	13.2	32.6	32.3	13.5	0.9	6.5	0.9
Great Britain	8.3	21.6	33.6	20.4	3.9	11.5	0.8
United States	8.6	17.2	31.8	20.1	5.2	16.9	0.3
Total	16.5	28.6	27.9	13.4	2.8	9.7	1.1

Notes: sa = strongly agree; a = agree; nn = neither agree nor disagree; d = disagree; sd = strongly disagree; Source: ISSP 2010, own calculations.

Treating missing values

As discussed above, the issue of missing data is not a salient problem of the ISSP 2010. Yet, missing values are encountered and the question arises of how to deal with them. For the variables that constitute general values and social positions, how missing values are treated depends on the scale level. For categorical variables which are transformed into dummy variables an additional dummy for the missing values will be created. For continuous variables the respective section discusses how missing values are dealt with (chapter 4.2.2). Due to the rather small amount of missing data, one could simply go ahead using list-wise deletion concerning the scales of environmental concern and action. However this would lead to a loss of some cases. In addition, the bootstrap option and the calculation of modification indices are not possible in AMOS when missing values are present. Thus it was decided to deal with this data in a somewhat unconventional way thus far not found in literature on the subject. MCA allows missing values to be included as an extra category. This will yield a graphical solution in which the missing values' categories are placed in relation to the meaningful categories. Thus, for each item, the missing value is replaced by the substantive category (for example the 'agree' category) to which it is closest instead of substituting it, for instance, with the scale mean.

4.2.2 The independent variables

As already mentioned in the introduction, environmentalism is a complex issue and consequently a vast amount of independent variables are discussed in theory and provided by the ISSP. The independent variables are operationalized as follows:

General values

Post-materialism is measured by asking which of the following statements is the most important for the country one lives in and which is the second most important.

1. Maintain order in the nation
2. Give people more say in government decisions
3. Fight rising prices
4. Protect freedom of speech

1 and 3 are considered materialist views, 2 and 4 post-materialist positions. If a post-materialist option was elected most important two points were given. A post-materialist option as second priority added one point. Zero points were

given for any materialist option chosen. The result is an index ranging from 0 (materialist) to 3 (post-materialist). This index is used for the macro analysis to compare the national means of post-materialism. There are very few missing values in these items and they are set on the median value. For the multivariate analysis two dummy variables are computed. The materialists are those holding a value of zero in the additive index and comprise the first dummy. Those with a value of one or two are the mixed types. The reference group is the post-materialist category which includes those with an index value of 3.

Some argue that only the long version of the post-materialism scale consisting of 16 items is able to reliably identify the relation between post-materialism and environmentalism (Davis 2000). The ISSP, instead, only provides the short four-item version which can be resorted to.

The *political position* is measured on a 5-point scale, 1 'Far left (communist, etc.)' to 5 'Far right (fascist, etc.)'. These are based on the country-specific party preferences. However, there are two more categories that disturb the left to right order and therefore pose some problems to the scale: 6 'other, no specification' 7 'no party affiliation, no preference, would not vote'. Additionally in some cases there is a relatively large amount of missing values due to those who refused to answer the question. To deal with this issue, dummy variables are created, one for each of the five basic categories and one for the non-voters combining category 6 and 7. The reference category in the multivariate analysis will be those who did not provide an answer or refused the question. Thus, not only the problem of missing values is dealt with but one can also answer the question if non-voters or 'question-refusers' are significantly different from those who stated their political position. Another problem with this scale arises in the USA. The respondents in the USA were only categorized into the three categories 2 (central, left), 3 (central), 4 (right, conservative) even though in the original USA questionnaire more option were available (see below for the original categories). Therefore, doubt rises that these three categories cover the political spectrum in the USA properly. This categorization might be due to the fact that in the USA a two-party system prevails. To handle this particularity, the original categories were used to build the dummy variables: strong democrat, not very strong democrat, independent (close democrat), independent, independent (close republican), not very strong republican, strong republican. The reference category includes those who did not answer or stated affiliation to another party. Due to the small number of respondents in these two categories they were combined.

To measure *generalized trust* four 5-digit-items are used comprising social and political trust.

1. Generally speaking, would you say that most people can be trusted, or that you can't be too careful in dealing with people? (1: You can't be too careful; 5: Most people can be trusted).
2. Generally speaking, would you think that most people would try to take advantage of you if they got the chance, or would they try to be fair? (1: Most people would try to take advantage; 5: Most people would try to be fair).
3. Most of the time we can trust people in governments to do what is right (1: strongly agree; 5: strongly disagree)
4. Most politicians are in politics only for what they can get out of it personally (1: strongly agree; 5: strongly disagree)

In items one, two and four higher values indicate more trust while in item three lower values entail more trust. To adjust item three to the other ones it was recoded so that a higher value indicated higher trust. A CatPCA showed that a one dimensional solution is feasible in all countries. The four items are added and divided by 20 (the maximum that can be reached) to achieve an index from 0 to 1. To deal with the relatively low amount of missing values in these items, all missing values were set to be equal to the middle category. This did not change the results of the multivariate analysis substantially in comparison to analysis using listwise deletion.

Social positions

Besides the fact that different currencies complicate the comparison of income or income effects, the *household income* is not asked for consistently. In Germany or Czech Republic respondents are requested to announce their monthly net household income while in Norway the gross household income per year is asked for and in the USA the household income per year before taxes is captured. To gain comparable results, first, the equivalence income is calculated by dividing the household income by the square root of the number of persons living in the household. Second, the equivalence income is z-standardized. This results in a mean of 0 and a standard deviation of 1 and produces a "country-specific measure of relative individual income" (Franzen and Meyer 2010, p. 227). To deal with missing values the following strategy is pursued: first, the deciles of the income variable were calculated. Second, the variable asking people to judge their socio-economic position on a scale ranging from 1=lowest to 10=highest was consulted.[8] If there was a missing value in the income variable but a valid

8 The variable in which the respondents have judged their socio-economic position is not used in the multivariate analysis. The self-judgment simply served as a reference to treat missing values in the income variable.

answer in the self-positioning scale, then the income was set on the decile equal to the category chosen in the self-positioning variable. For example, if a respondent stated he or she belonged to category 4 than the household income for that person was set to the fourth decile. In case of a missing value in both variables, the income was set to be equal to the median income.

To capture the respondent's *educational attainment*, they are grouped into the following categories:

'No formal education'; 'Lowest formal qualification'; 'Intermediate secondary completed'; 'Higher secondary completed (usually entry requirements for universities)'; 'University degree incomplete'; 'University degree complete'

For each category a dummy variable is created with the 'No formal education' option being the reference category. An additional dummy variable for those who did not give an answer will be included to handle missing values.

The biographic availability assumption is represented by two aspects: the respondent's *employment situation* and the *number of children living in the household*. The employment situation is measured by a categorical variable with the categories 'never had paid work', 'currently not in paid work' and 'currently in paid work'. In the multivariate analysis a dummy variable for each category ('never had paid work' serves as reference category) and one for those who did not provide an answer will be included. The number of children living in a household is more delicate because the respondents are asked for the number of toddlers and the number of schoolchildren. The two variables are included as continuous variables. In this case, missing values are substituted by setting them to the median value which is zero across all countries.

Finally, the *size of the community* the respondents live in is represented by a categorical variable: 'big city', 'suburbs of a big city', 'town or small city', 'village', and 'farm in the country'. 'Big city' serves as the reference category and for each other category a dummy variable will be used including one for missing values.

4.2.3 The dependent variables

The salience of the environment

For the country-specific profile, a descriptive introduction to the salience of the environment and environmental problems will be the starting point. Three topics are captured by the ISSP data: the environment in relation to other public issues, various environmental problems in relation to each other, and future energy supply. Two items (Q1a and Q1b) measure the relation of the environment to other public issues by asking the respondents to state which issue is most important for the country and which one is second most important (1: health care; 2: education; 3: crime; 4: the environment; 5: immigration; 6: the economy; 7: terrorism; 8: poverty; 9: none of these). In the tables, the absolute frequencies of each category are presented and a score for each category was calculated:

Score = frequency of most important+frequency of second most important ÷ 2.

For instance: if the environment was chosen 10 times as most important and 18 times second most important, score = 10 + 18 ÷ 2 = 19.

Q7a and Q7b request respondents to judge which environmental problem is most important for the country and which one is most important for oneself and the family, respectively (1: air pollution; 2: chemicals and pesticides; 3: water shortage; 4: water pollution; 5: nuclear waste; 6: domestic waste disposal; 7: climate change; 8: genetically modified foods; 9: using up our natural resources; 10: none of these). A score is also calculated for the environmental problems in relation to each other, analogous to the formula above, with one exception. The importance for the country and for the family are weighted equally, so:

score = most important for country+most important for family.

Finally, the respondents had the possibility to state the best way in which to meet the country's future energy needs (Q18) (1: coal, oil, and natural gas; 2: nuclear power; 3: solar, wind or water power; 4: fuels made from crops; 5: none of them; 6: USA only: fuels made from waste). The relative frequencies will be provided in the tables.

Approaching the environment via ranking items in relation to other issues might yield additional information that are maybe concealed when only regarding the absolute environmental concern measured via rating scales. For instance, people's willingness to pay for environmental protection in one country might be quite high but when comparing the environment to other issues it is possible that many other things are more important. This way of examination can help to explain why, as suggested by the low-cost hypothesis, the high willingness to pay does not transfer into action: other topics have priority over the environment.

Three dimensions of environmental concern

The secular environmental attitudes and behaviors form a complex and versatile construct (Preisendörfer 1996a, p. 3). A wide range of questions concerning environmental attitudes is asked in the ISSP. They are necessary to capture attitudes and behavior as close to reality as possible. These questions direct different dimensions of environmental attitudes which are briefly described in the following section.

As mentioned in 3.3.1, three dimensions of environmental concern (cognitive, evaluative and conative) are discussed in literature on the subject. From a theoretical as well as an analytical perspective, it makes sense to separate them because then a test can be conducted to determine how far the cognition leads to evaluative reactions, and from there to behavioral intentions and eventually to action itself. Additionally the researcher can test which of these dimensions is directly affected, for instance, by education.

Concerning the cognitive and evaluative dimension Lachmann, (forthcoming) evaluated a scale used for instance by Franzen and Meyer (2010). The results of the MCA clearly indicate that this combination of items is a poor measure of environmental concern in most nations taking part in the ISSP 2010.[9] Hence, for the present study these two dimensions are represented differently, as the following sections will illustrate.

The cognitive dimension

The perceived environmental threats represent the cognitive dimension of environmental concern in the present study. Resolved from objective environmental problems, the subjective perception of environmental threats is hypothesized to positively influence attitudes and action.

Several questions address the subjective perception of environmental threats, asking if the given issue is 1: extremely dangerous for the environment; 2: very dangerous 3: somewhat dangerous; 4: not very dangerous; 5: not dangerous at all, for the respective country:

- Q14a In general, do you think that **air pollution caused by cars** is…
- Q14b In general, do you think that **air pollution caused by industry** is …
- Q14c And do you think that **pesticides and chemicals used in farming** are …
- Q14d And do you think that **pollution of COUNTRY'S rivers, lakes and streams** is…
- Q14e In general, do you think that **a rise in the world's temperature caused by climate change** is …

9 Following items were used: Q9c Modern science will solve our environmental problems with little change to our way of life; Q10a We worry too much about the future of the environment and not enough about prices and jobs today; Q10c People worry too much about human progress harming the environment; in order to protect the environment (country) needs economic growth; Q13a It is just too difficult for someone like me to do much about the environment.

As mentioned in chapter 4.2.1, Q14f and Q14g are dropped because of the relatively high number of missing values.

The evaluative dimension

Most problematic in finding variables measuring environmental concern was the identification of items that might adequately represent the evaluation of environmental issues. Six items with most appropriate content were chosen:

- Q10a We worry too much about the future of the environment and no enough about prices and jobs today
- Q10c People worry too much about human progress harming the environment
- Q13a It is just too difficult for someone like me to do much about the environment
- Q13c There are more important things to do in life than protect the environment
- Q13d There is no point in doing what I can for the environment unless others do the same
- Q13e Many claims about environmental threats are exaggerated

The application of MCA (as well as Principal Component Analysis and Categorical Principal Component Analysis) yields results that strongly vary between countries with regard to dimensionality. As this is a cross-national comparative study, measures for this dimension are needed that are comparable. To achieve comparability a common core of the evaluative environmental concern needed to be found. For that reason a simple Principal Component Analysis (PCA) was conducted for each nation. In each country three items with highest factor loadings on the first dimension were chosen to represent the evaluative dimension of concern in that country. Three items were selected because the first dimension comprises at least three items in every nation with factor loadings of $f > 0.6$. This results in a country specific measure of the evaluative environmental concern but the size of the correlation/covariance matrix is equally large and factor loadings are relatively close to each other. Table A2 in the appendix lists which items form this dimension in which country.

The conative dimension

To represent the conative dimension of environmental concern, the respondents were asked if they would pay much higher prices (Q12a), pay much higher taxes (Q12b) or accept cuts in the standard of living (Q12c) to protect the environment. This dimension is the most frequently used scale of environmental concern and Lachmann (forthcoming) found it to be of good quality in most nations. However, in Chile, Latvia, Israel (Arabs), South Korea, Slovak Republic, and Taiwan the scale does not work well as some abnormalities in MCA were encountered. The extent of abnormalities in the data concerning the other dimensions of environmental concern will be answered in the next section.

Multiple Correspondence Analysis (MCA) was applied to assess the quality and especially the comparability of the cognitive and evaluative scale of environmental concern. As stated above the conative dimension is not problematic (with some exceptions), hence, only the evaluative and cognitive dimension will undergo further investigation. As Table 4.4 clearly shows, despite the reduction to a very narrow representation comprising three items, the evaluative scale does not work well in each sample.[10] For Argentina, Chile and the Philippines the two dimensions changed, meaning that the first dimension represents the methodologically induced variance while the second dimension captures the substantive variation.

10 In general three items may be too few for an extensive evaluation of the scale via MCA. However, as this three item approach yielded the best comparable scale, the evaluation via MCA is nevertheless taken to get a basic insight into the functioning of the scale.

Table 4.4: MCA of the evaluative dimension of environmental concern

	Variance	Pattern	Category interchanges
Argentina	57.6	D1 D2 changed	No changes
Austria	64.4	OK	No changes
Belgium	61.3	OK	No changes
Bulgaria	64.9	OK	No changes
Canada	67.7	OK	No changes
Chile	67.5	D1 D2 changed	7
Taiwan	48.1	Very unsystematic	Uncountable
Croatia	61.8	OK	No changes
Czech Republic	58.3	Between D1 D2	No changes
Denmark	66.7	OK	No changes
Finland	61.4	OK	No changes
France	63.8	OK	No changes
Germany West	58.9	OK	No changes
Germany East	56.8	OK	No changes
Israel Jews	60.3	OK	3
Israel Arabs	65.2	Very unsystematic	Small sample size
Japan	57.9	Between D1 D2	No changes
South Korea	56.7	Between D1 D2	3
Latvia	61.5	OK	4
Lithuania	60.8	OK	2
New Zealand	67.9	OK	No changes
Norway	66.7	OK	No changes
Philippines	61.1	D1 D2 changed	No changes
Slovak Republic	62.4	OK	No changes
Slovenia	58.7	OK	No changes
Spain	52.3	OK, slightly leaned to one side	1
Sweden	63.5	OK	No changes
Switzerland	55.3	Between D1 D2	No changes
Great Britain	69.1	OK	No changes
USA	63.1	OK	No changes

Notes: In the column Pattern 'D1 D2 changed' means that the order of the categories is mirrored on dimension 2 while the artificial variation is represented on dimension one. Source: ISSP 2010, own calculations.

In these three countries more variation is due to methodological artifacts than to substantive responses. Additionally, in Chile several misplaced categories were found, again an indication that data quality is not good in that country. Other samples in which the results question the data quality are Taiwan, Israel (Jews and Arabs), South Korea as well as Latvia and Lithuania. In Taiwan and Israel (Arabs) a very unsystematic pattern was found with the result of many violations of the ordinality of the categories. For Israel (Jews), South Korea, Latvia and Lithuania the categories of the three items form a horseshoe in general but had misplaced categories in two or more items. Figure 4 gives an example of a country in which the scale works well (New Zealand) and one in which problems occurred (Chile).

In Czech Republic, Japan and Switzerland the column 'Pattern' says 'Between D1 D2'. In these three countries the categories form a horseshoe, which indicates good data but it is slightly turned so that the order of the categories lies between the two dimensions but their order is retained. This may be an indication that a two dimensional solution might be better than a single dimension. Below, the structural equation model will give a deeper insight into this issue.

Figure 4: Cognitive maps of the evaluative dimension of environmental concern: two examples

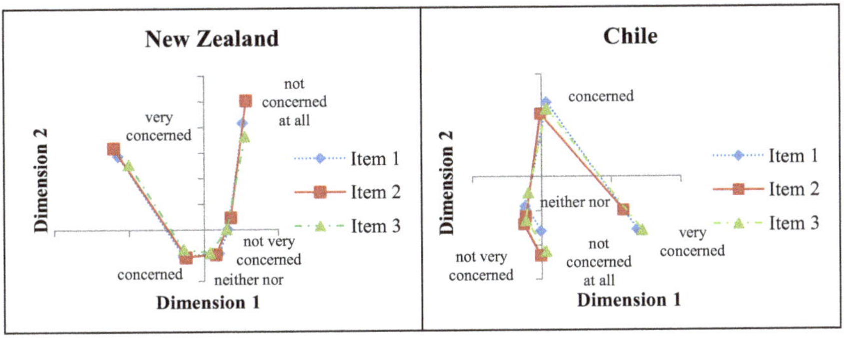

Next, the MCA was applied to assess the cognitive dimensions of environmental concern. Q14a-Q14g (see above for wording) directly asked to what extent various issues are dangerous for the environment. In general, the 5th category is used very infrequently, in some instances not at all. The 4th category, in most cases, was chosen much less frequently than the rest as well. Thus, a skew of the graphical representation is expected to show itself as well as interchanges between these two categories. The analysis was rerun with the 5th and the 4th category combined into one single category. Beforehand, item Q14f and Q14g are excluded from the analysis because they hold a substantially higher number of missing values (see chapter 4.2.1).

In total, two MCAs are run for every nation:

1) Q14a-e
2) Q14a-e categories 4 and 5 combined

The results are summarized in Table 4.5. Figure 5 gives two examples: Czech Republic where no problems occurred and Israel (Jews) where even after merging category 4 and 5 interchanges were still found.

Table 4.5: MCA of the cognitive dimension of environmental concern

	Variance	Variance 4 and 5 combined	Pattern	Pattern 4 and 5 combined	Category interchanges	Category interchanges 4 and 5 combined
Argentina	56.9	56.7	OK	OK	2	No changes
Austria	50.0	50.0	OK	OK	1	1
Belgium	54.0	53.9	OK	OK	No changes	No changes
Bulgaria	53.9	53.8	Between D1 D2	OK	3	2
Canada	57.5	57.3	OK	OK	No changes	No changes
Chile	61.7	61.6	D1 D2 changed	D1 D2 changed,	4	1
Taiwan	50.4	50.4	OK	OK	5	No changes
Croatia	59.5	59.4	OK	OK	No changes	No changes
Czech Republic	54.8	54.4	OK	OK	No changes	No changes
Denmark	55.2	55.1	OK	OK	5	No changes
Finland	53.5	53.7	OK	OK	4	No changes
France	51.9	51.9	OK	OK	No changes	No changes
Germany West	53.7	53.7	OK	OK	3	No changes
Germany East	49.8	49.2	OK	OK	No changes	3
Israel Jews	49.0	49.0	OK	OK	4	3
Israel Arabs	48.8	48.7	Almost OK	Between D1 D2	5	No changes
Japan	58.8	56.6	OK	OK	No changes	No changes
South Korea	60.0	56.8	D1 D2 changed	OK	17	No changes

	Variance	Variance 4 and 5 combined	Pattern	Pattern 4 and 5 combined	Category interchanges	Category interchanges 4 and 5 combined
Latvia	60.1	49.3	OK	OK	No changes	No changes
Lithuania	57.4	57.2	OK	OK	No changes	No changes
New Zealand	55.9	55.8	OK	OK	2	No changes
Norway	54.9	51.7	OK	OK	No changes	No changes
Phillipines	50.8	50.4	OK	OK	1	1
Slovak Republic	57.6	57.5	OK	OK	No changes	No changes
Slovenia	58.0	50.6	D1 D2 changed	OK	20	1
Spain	53.8	53.7	OK	OK	3	No changes
Sweden	56.7	56.3	OK	OK	4	No changes
Switzerland	50.5	50.4	OK	OK	No changes	No changes
Great Britain	58.9	58.9	OK	OK	4	No changes
USA	59.5	59.4	OK	OK	7	No changes

Notes: In the column Pattern and Pattern 4 and 5 combined 'D1 D2 changed' means that the order of the categories is mirrored on dimension 2 while the artificial variation is represented on dimension one. Source: ISSP 2010, own calculations.

After combining categories 4 and 5, the situation became better in almost all countries. Only in eastern Germany, the 'uncombined' scale yields better results. In Chile, Israel (Arabs), and Philippines neither of the two variants offer a satisfying result. Concerning the explained variance (column 2 and three in Table 4.5), the decline when combining 4 and 5 is not surprising because of the loss of information caused by the combination. Yet the decline of explained variance is quite small. In total, the merging of the two categories seems justified and the scale should rather be used in that way (with exception for eastern Germany where the full scale can be used).

Figure 5: Cognitive maps of the cognitive dimension of environmental concern: two examples

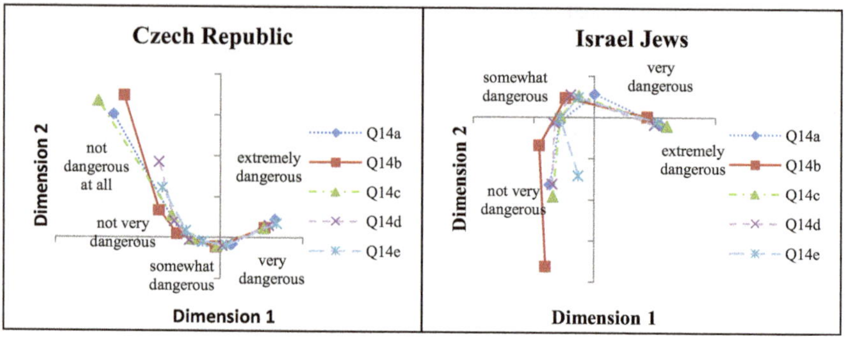

Environmental action

So far different aspects of environmental concern have been analyzed. However, environmental action is of major interest. The items measuring this underwent the same evaluation as the other items in the previous part. In terms of the socio-economic, political and legal actor, three types of action are considered for the present study. First, the day-to-day routine in which individuals may or may not act in an environmentally friendly way is addressed: the socio-economic actor. Because such behavioral adjustments need changes in daily activities, collection of prior information, higher prices, and additional time spent, etc., this is a medium-cost activity. Second, the acceptance of laws will be operationalized, representing the legal actor. Actually, this aspect captures non-action rather than action. Responsibility is transferred to the government instead of oneself and if there are behavioral changes to the individual, these are prescribed and people do not have to make an effort to find the right thing to do. This represents a low-cost 'activity'. However, people can have an influence on environmental legislation. Either by voting for parties with a strong environmental profile or as the political actor illustrates, by the third aspect of pro-environmental action: environmental activism. This is the high-cost activity, as it demands most resources from people, from time needed to demonstrate, money to donate and finally the effort to gather information about an appropriate organization that one is willing to support. The three aspects are operationalized as follows.

Everyday activities

In particular there are six items asking the respondents to state if they 'always' (A), 'often' (O), 'sometimes' (S) or 'never' (N) conduct given activities. This is the dimension of everyday action. Table 4.6 provides the item numbers, wording of the items, the marginal distribution plus the percentage of missing values.

Before applying MCA to the scale of environmental action, a look at the marginal distribution indicates that Q20a, Q20b and Q20c should not be included for the following reason: in all three cases more missing values are found (Table 4.6). Especially in less-developed nations, more respondents state that recycling, untreated vegetables or fruits or driving a car are not available there (results not shown). For instance, 33.5%, 34.1%, and 31.0% of the respondents in Bulgaria, Latvia and Russia, respectively, answered that recycling is not available. In contrast, in Switzerland, Austria, and Belgium only 0.0%, 0.1% and 0.2%, respectively, have no access to recycling. In addition, concerning Q20a, in developed nations people are much more engaged in recycling than in the other behaviors. These three items are not only constrained by people's ambition but also by contextual factors (availability) while the other three are readily available to everyone.

Table 4.6: Everyday action: relative frequencies

ID	Wording	A	O	S	N	Missings
Q20a	How often do you make a special effort to sort glass or tins or plastic or newspapers and so on for recycling?	43.3	20.6	16.1	12.5	7.6
Q20b	How often do you make a special effort to buy fruits and vegetables grown without pesticides or chemicals?	10.6	23.3	31.9	25.7	8.5
Q20c	And how often do you cut back on driving a car for environmental reasons?	4.1	13.4	27.5	29.3	25.7
Q20d	How often do you reduce the energy or fuel you use at home for environmental reasons?	13.1	28.8	32.9	23.3	1.8
Q20e	And how often do you choose to save water or re-use water for environmental reasons?	13.7	21.6	31.2	27.6	1.5
Q20f	And how often do you avoid buying certain products for environmental reasons?	8.4	24.0	37.8	27.3	2.5

Notes: Q20a, Q20b and Q20c have the additional category staing that recycling, untreated fruits/vegetables or a car is not available. These categories account for the majority of missing values.
Source: ISSP 2010, own calculations.

The scale comprising items Q20d-f works quite well in most nations. In countries that caused problems regarding the scales measuring environmental concern, similar problems appear within this scale. Bulgaria, Israel (Arabs), Latvia, and Philippines, again yield unsatisfactory results, either due to an exchange of the substantive and methodological dimension (Bulgaria), an unsystematic response pattern (Israel), various interchanged categories (Latvia) or a strong skew accompanied by category interchanges (the Philippines). Categories interchanges also appear in Bulgaria, Israel, and Latvia.

Acceptance of environmental laws

Two items ask the respondents to state their acceptance of environmental laws. They had to choose which of the statements were closest to their own opinion:

- Q15a Government should let **ordinary people** decide for themselves how to protect the environment, even if it means they don't always do the right thing or Government should pass laws to make **ordinary people** protect the environment, even if it interferes with people's rights to make their own decision.
- Q15b Government should let **businesses** decide for themselves how to protect the environment, even if it means they don't always do the right thing or Government should pass laws to make **businesses** protect the environment, even if it interferes with people's rights to make their own decision.

A MCA with the two items was applied to test if both belong to the same dimension. In this step the factor scores for each respondent were saved as a latent variable (with a mean of zero and a standard deviation of one) which is used for the multivariate analysis. Missing values were set to the value they were located closest to in the MCA analysis. Across all countries this was the category where the respondents stated that people/businesses should decide themselves how to protect the environment.

Environmental activism

To measure people's degree of political environmental activism four items ask if the respondent

- is a member of any group whose main aim is to preserve or protect the environment

and if in the past five years he or she

- signed a petition about an environmental issue
- gave money to an environmental group
- took part in a protest or demonstration about an environmental issue

The MCA yields a one-dimensional solution. Due to the very small amount of missing values these are set to the modus, i.e. answering 'no' to the respective question. In the same way as for the acceptance of laws, the factor scores for each respondent were saved as a new variable, again with a mean of zero and a standard deviation of one. This latent variable, then, will be used in the multivariate analysis.

After this inspection of data quality, a decision has to be made on which countries provide data that is good and sufficiently comparable to undergo substantive analysis. These are: Austria, Belgium (Flanders), Canada, Croatia, Czech Republic, Denmark, Finland, France, Germany (East and West), Japan, New Zealand, Norway, Slovenia, Spain, Sweden, Switzerland, Great Britain, and the USA. In total 19 samples remain for analysis. Mexico, Russia, Turkey, and South Africa are not included due to a relatively high amount of duplicated interviews, for instance when an interviewer conducted several interviews and then copied and pasted some of them with minor changes (Lachmann forthcoming). The other country samples are not used due to irregularities in the scales measuring environmental concern and action as presented above.

4.2.4 Response pattern biases

Up to now the cognitive maps of the various scales of environmental concern and action by country have been evaluated. The graphical visualizations show the response pattern or the skew of the distribution accounts for some of the methodological variance. Typical causes of methodologically induced variance are social desirability and response pattern biases. Blasius and Thiessen (Blasius and Thiessen 2012, p. 6) speak of "normative demands and impression management" when they address responses that represent three images that one has of oneself: how does one think of oneself, how does someone else think of oneself, and how would one like others to think? Are the answers provided substantive or biased by some degree of social desirability/impression management?

Besides this kind of normative bias, task simplification can be an issue as well. Is there a tendency to agree (acquiescence) regardless of the meaning? Can an extreme response style be found (i.e. choosing SA or SD) or is it mostly the midpoint category that respondents choose (Blasius and Thiessen 2012, p. 9)?

To get an insight into the response style, it is counted how often a person 'agrees' or 'strongly agrees', how many 'neither agree nor disagree' responses are provided and then how often 'disagree' or 'strongly disagree' answers are given; these are counted separately for every category. 26 items in the ISSP ask for agreement or disagreement. After counting how often the categories are chosen, the number is divided by 26, i.e. the number of items. This provides a ratio of how many of the responses fall into the respective categories. Figure 6 presents the ratios by country.

All in all, the amount of agreement and disagreement is relatively balanced and there is definitely no tendency to give NN responses. In none of the 19 countries does the agreement rate reaches 0.45. The highest rates can be found in Spain (0.44) and Slovenia (0.42). Canada holds the highest ratio of disagreement (0.42) followed by Norway (0.39). With regard to the NN category, Belgium (Flanders) ranks first (0.31) and Sweden second (0.29). In Austria, Canada, Finland, New Zealand, Norway, Sweden, and the USA there are slightly more disagreeing answers than agreeing ones.

Figure 6: Response patterns

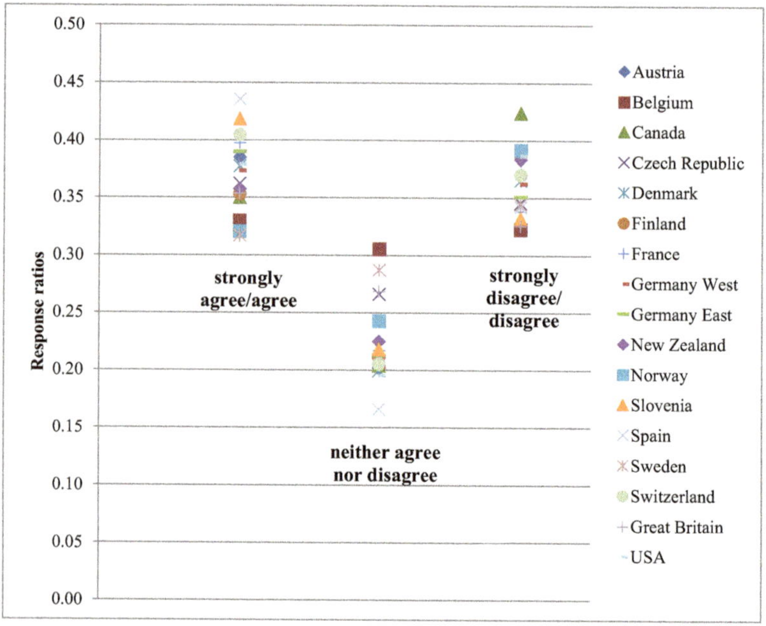

Due to the fact that six of these 26 items have no environmental content (Q2a, Q2b, Q5a Q5b, Q9a, Q9b), and that in 9 items (Q10b, Q11b-c, Q12a-c, Q13g, Q19a) agreement indicates environmental concern, in 6 items (Q10a, Q10c, Q13a, Q13c-e) disagreement indicates environmental concern and in 5 items (Q9c, Q11a, Q13f, Q19b-c) it is not absolutely clear whether agreement or disagreement means more environmental concern, it seems that no serious issue of task simplification or the tendency to provide social desirable answers is present.

5. A global environmental profile

Before going into a detailed investigation of environmentalism on the micro level, a global profile will be created. Global, in this case, is of course restricted to nations with sufficiently high and comparable data quality. The countries are spread around the world but most of them are European countries. Even though there is a substantial amount of variation in the GDP per capita, all of them would qualify as high-income countries (see www.worldbank.org). Descriptive statistics will provide a comparative insight into the status quo. Response ratios are calculated and presented for the descriptive statistics. For each dimension of environmental concern and action, it is counted how often environmentally friendly, neutral and environmentally unfriendly answers are provided, each then divided by the total number of responses given in the respective set of variables (see Chapter 4.2.3 for individual variables). The ratio of environmentally friendly answers will be the country value for the macro analysis.

Table 5.1: Descriptive statistics on the macro level

	GDP per capita[a]	Post-materialism[b]	Cognitive[c]	Evaluative[c]	Conative[c]	Everyday Action[c]	Laws[c]	Activism[c]
Austria	44916	1.24	0.63	0.57	0.34	0.46	0.79	0.14
Belgium	43006	1.04	0.51	0.43	0.30	0.48	0.71	0.12
Canada	46212	1.48	0.62	0.54	0.36	0.48	0.75	0.16
Croatia	13461	1.07	0.72	0.38	0.15	0.29	0.78	0.06
Czech Republic	18910	1.00	0.65	0.36	0.14	0.37	0.71	0.07
Denmark	56486	1.46	0.55	0.49	0.38	0.43	0.82	0.12
Finland	43864	1.54	0.51	0.58	0.29	0.43	0.64	0.12
France	39170	1.19	0.58	0.38	0.25	0.62	0.72	0.14
Germany East	40164	1.72	0.67	0.51	0.29	0.49	0.78	0.08
Germany West	40164	1.76	0.72	0.50	0.34	0.52	0.72	0.13
Great Britain	36256	1.31	0.40	0.33	0.23	0.34	0.66	0.11
Japan	43063	1.36	0.65	0.39	0.29	0.48	0.86	0.05
New Zealand	32407	1.16	0.49	0.38	0.34	0.35	0.71	0.19
Norway	85443	1.19	0.37	0.41	0.32	0.27	0.78	0.11
Slovenia	22898	1.02	0.66	0.46	0.25	0.39	0.86	0.06
Spain	29956	1.16	0.75	0.48	0.26	0.43	0.75	0.09
Sweden	49360	1.35	0.53	0.48	0.31	0.29	0.81	0.12
Switzerland	70561	1.17	0.52	0.54	0.53	0.49	0.59	0.19
USA	46612	1.51	0.55	0.41	0.37	0.34	0.59	0.11
Total	42258.37	1.30	0.59	0.45	0.30	0.42	0.73	0.11

Notes: [a] current US$; [b] country mean; [c] ratio of environmental friendly answers. Source: ISSP 2010, own calculations.

Table 5.1 depicts the ratios of environmentally friendly answers for the three dimensions of environmental concern and action differentiated by country. Additionally, the GDP per capita and the nations' average post-materialism are shown. One initial observation is declining environmental concern from one dimension to the next. Where the global ratio of environmentally friendly answers of the cognitive dimension is 0.59, the evaluative reactions reach only 0.45 and the willingness to do something declines even to 0.30. Furthermore, there is not only variation between the dimensions but also within them. Whereas Spain's cognition of environmental threats is 0.75, in Norway it is 0.38 points lower. Finnish respondents exhibit strongest evaluative reactions (0.58), which decline to 0.33 in Great Britain. The differences found in the conative dimension are equal to those in the cognitive dimension. Swiss people are much more willing to do something (0.53) than those in Czech Republic (0.14), a difference of 0.39 points.

In their article "The wealth of nations and environmental concern", Diekmann and Franzen (1999) found a positive relationship between the nations' GDP per capita and their environmental concern using ISSP data of 1993. Correlating the ratios of environmental concern with the GDP per capita of the 19 countries, shows in positive relationship in the evaluative and conative dimension. A negative coefficient is present in the cognitive component for the present study. The correlations are relatively large: $r_{cognitive}=-0.612$ $p<0.01$; $r_{evaluative}=0.319$, n.s.; $r_{conative}=0.730$, $p=0.001$. It has to be noted that there are only 19 countries in the analysis which definitely reduces the significance level, while outliers might have an influence on the effects size. It could be concluded from this finding that the wealthiest (of the wealthy) nations are more environmentally concerned and much more willing to make sacrifices. In contrary, the less wealthy of these 19 relatively rich countries perceive environmental threats more strongly. The last statement would go along with the assumption that objective problems are more pronounced in less wealthy nations.

What about environmental action? Table 5.1 also presents the ratios of environmental everyday action, the acceptance of laws and activism. If interpreted as a percentage, in total 42% of the respondents are always or often engaged in everyday action, in France 62% while in Norway only 27%. This is a difference of 35 percentage points. The correlation of GDP per capita and everyday action is very low and not significant ($r=0.014$, n.s). The acceptance of environmental laws (column 'laws' in Table 5.1) holds the highest ratios of the three aspects of environmental action. In Japan and Slovenia people most strongly prefer legal regulations over voluntary environmental protection (both 0.86), while in Switzerland and the USA this ratio is lowest (both 0.59). The correlation with GDP is small and not significant (r=–0.151; n.s.). In contrast to the other measures of environmentalism, the ratios of environmental activism are very low. In total, if interpreted as a percentage, only 11% of the respondents were active in the past five years, with substantive between-country variation. Whereas in New Zealand or Switzerland 19% participated in activism, this figure was only 5% in Japan. This high-cost activity positively correlates with GDP per capita (r=0.452; p=0.052).

Besides the fact that GDP is not a very good indicator of a nation's wealth (Van den Bergh 2007), correlations, especially with such a small sample size, can be misleading as well. As correlations express a linear relationship, other associations between variables can easily be concealed by the coefficient. Hence, an inspection of scatterplots might yield further insights.

Figure 7 shows the scatterplots of GDP per capita and post-materialism in relation to environmental concern. With decreasing GDP per capita the perception of environmental threats increases linearly. In the present sample people in less wealthy nations are more sensitive to environmental problems, maybe because these problems are more salient. The relation to the evaluative and conative dimension is the other way round, however, in a nonlinear manner. Richer countries exhibit higher ratios of environmentally friendly answers. As illustrated by the lines, the relation between GDP and the conative dimension seems to level off after a certain degree of wealth is reached. With regard to the evaluative dimension after a turning point, further wealth decreases people's evaluative reaction. This tendency still holds when excluding the riches country in the sample, Norway. This may be explained in terms of the environmental attitude's Kuznets curve: after a certain degree of wealth, environmental issues become more salient and part of everyday life and thus emotional and normative reactions are less pronounced.

Post-materialism is not related to the perception of environmental threats. Concerning the conative dimension, there is a similar tendency to that relevant to GDP per capita. First the value increases with increasing post-materialism and, after a turning point is reached, the willingness to pay begins to decrease again. For the evaluative dimension, the relation is almost linear but, as the R^2 value shows, not overly strong.

With this very restricted data set the assumption of the post-materialism hypotheses cannot be supported fully. One of the dimensions (cognitive) is not related to post-materialism at all; one relation is quite linear (evaluative), the other rather in form of an inverted U-shaped curve (conative). The link between wealth and environmental concern is also less straightforward than the wealth and affluence hypothesis would predict. Two non-linear curves are discovered and one linear negative one. However, the results show that a nation's wealth as well as post-materialism are related to environmental concern. This implies that within this sample the globalization hypothesis does not find support regarding environmental concern.

Figure 7: Scatterplots – GDP per capita, post-materialism and environmental concern

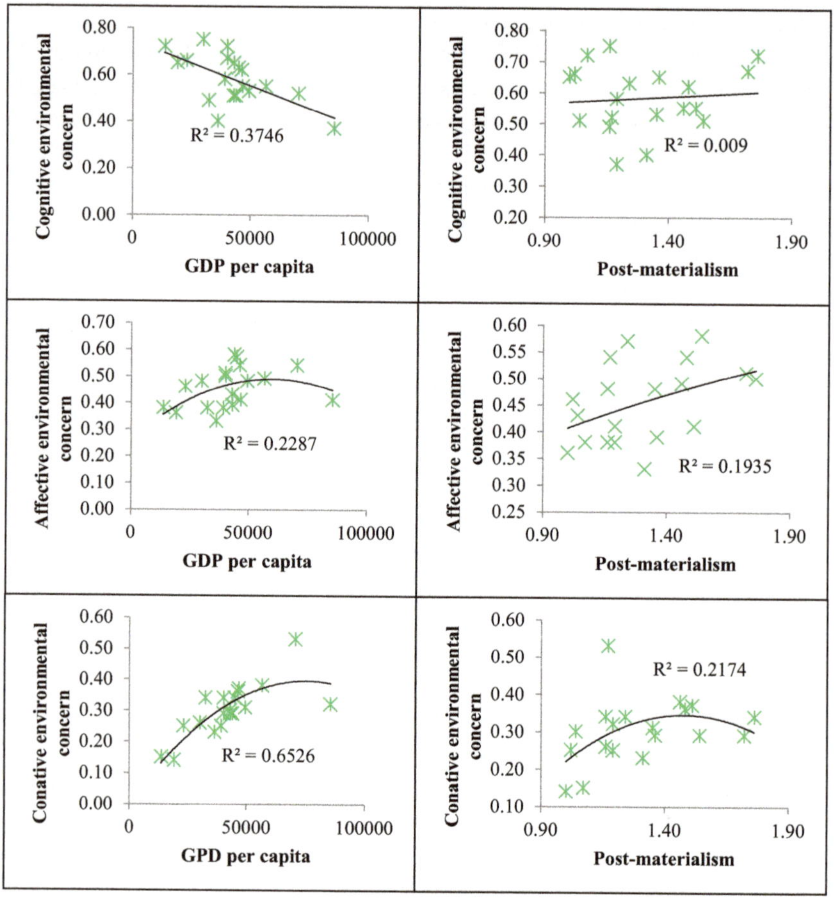

Having a look at Figure 8, neither environmental everyday action nor the acceptance of laws are related to a nation's GDP per capita, nor to the average degree of post-materialism. Only activism is be positively correlated with GDP per capita. The concave trajectory of activism and post-materialism implies a small nonlinear relationship. The low R^2 clarifies that these two categories are at best marginally connected.

Figure 8: Scatterplots: GDP per capita, post-materialism and environmental action

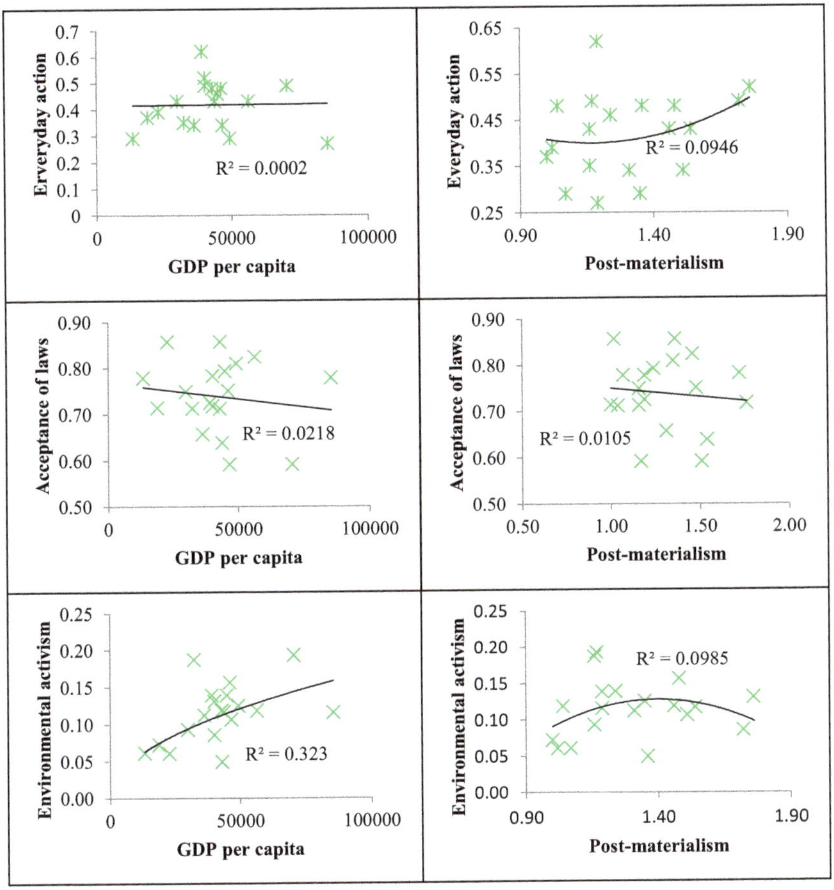

In the next step, the first insight into the nationspecific relationship between environmental attitudes and actions will be provided before six detailed country profiles will be created. To start with, a simple path analysis model is made up proposing causal relationships between the different dimensions of environmental concern and environmental action. The path diagram was already presented in Figure 2. As modern structural equation modeling software allows multi-group analysis, this results in regression coefficient and explained variances (R^2) for each country, which in this regard is an advantage over a random slope and random intercept model in a multilevel framework (see chapter 4.1). Thus, the effects within different countries can be compared.

As Table 5.2 illustrates, the extent to which pro-environmental everyday action can be explained by pro-environmental attitudes varies substantially between the 19 countries. The way in which environmental concern contributes to everyday action is quite versatile. It is not only the amount of variance explained by environmental concern that is different. Mostly a mixture of value and instrumental rational considerations induces people to act in environmentally friendly ways.

In Belgium as well as in Germany West, only 6.7% and 9.3% respectively of everyday action are attributed to environmental concern, while in New Zealand, Great Britain, Canada and Finland this percentage is over 25%. It is also remarkable that the three different dimensions contribute quite differently to environmental action depending on the country. The strongest prediction by the cognitive dimension can be found in Slovenia (b=0.527) whereas in Germany West (b=0.076) the perception of environmental threats does not contribute significantly to environmental action at all. With b=0.396, the evaluative concern has its strongest effect in Finland. The lowest, non-significant effect is found in Germany East (b=–0.063). More surprising is the significantly negative coefficient in Japan (b=–0.148). A higher evaluative concern entails less engagement in everyday action, maybe as a result of some kind of resignation or country-specific mentality. However, both explanations are speculative. In Sweden (b=0.315) the impact of people's willingness to do something is strongest while in Croatia (b=0.022) it does not contribute to the explanation of everyday action.

Table 5.2: Explained variances of environmental action by environmental concern

	R^2: Everyday action	Cognitive --> Action	Evaluative --> Action	Conative --> Action
Austria	0.214	0.351* (0.206/0.512)	0.260* (0.129/0.405)	0.271* (0.138/0.428)
Belgium	0.060	0.147* (0.013/0.281)	0.090 (−0.066/0.247)	0.228* (0.097/0.355)
Canada	0.286	0.423* (0.325/0.544)	0.038 (−0.060/0.140)	0.174* (0.073/0.268)
Croatia	0.191	0.370* (0.278/0.467)	0.303* (0.178/0.441)	0.022 (−0.058/0.101)
Czech Republic	0.100	0.297* (0.193/0.406)	0.068 (−0.040/0.193)	0.162* (0.080/0.247)
Denmark	0.207	0.284* (0.165/0.415)	0.070 (−0.026/0.159)	0.238* (0.150/0.333)
Finland	0.254	0.095 (−0.022/0.216)	0.396* (0.270/0.530)	0.186* (0.079/0.294)
France	0.159	0.349* (0.266/0.444)	0.094* (0.008/0.178)	0.124* (0.056/0.190)
Germany-West	0.090	0.076 (−0.045/0.198)	0.455* (0.288/0.672)	0.031 (−0.117/0.169)
Germany-East	0.197	1.009* (0.490/2.403)	-0.063 (−0.373/0.184)	0.311* (0.139/0.478)
Japan	0.112	0.324* (0.213/0.427)	-0.148* (−0.259/−0.035)	0.312* (0.217/0.400)
New Zealand	0.335	0.323* (0.237/0.415)	0.056 (−0.009/0.123)	0.255* (0.170/0.354)
Norway	0.232	0.226* (0.117/0.337)	0.073 (−0.052/0.192)	0.214* (0.128/0.310)
Slovenia	0.237	0.527* (0.411/0.657)	0.110* (0.002/0.219)	0.023 (−0.063/0.103)
Spain	0.161	0.229* (0.160/0.303)	0.197* (0.120/0.285)	0.077* (0.028/0.128)
Sweden	0.226	0.193* (0.094/0.296)	0.031 (−0.067/0.138)	0.315* (0.229/0.416)
Switzerland	0.154	0.142 (−0.004/0.283)	0.295* (0.181/0.453)	0.183* (0.061/0.312)
Great Britain	0.360	0.400* (0.297/0.506)	-0.028 (−0.125/0.060)	0.278* (0.196/0.362)
United States	0.243	0.227* (0.153/0.309)	0.052 (−0.032/0.142)	0.235* (0.160/0.319)
Total	0.168	0.292* (0.268/0.316)	0.129* (0.102/0.152)	0.172* (0.152/0.194)

Notes: Model fit: CFI=0.931; RMSEA=0.015; SRMR=0.056; unstandardized coefficients; confidence intervals in parentheses.
Source: ISSP 2010, own calculations.

In general, the evaluative dimension seems to be the weakest of the three predictors, as a significant coefficient was found in only seven countries. With regard to the other two dimensions, there are only three countries where these effects are not significant. The perception of environmental threats is the strongest predictor because it carries larger coefficients than the conative or the evaluative environmental concern. Environmental everyday action seems to be more instrumental than value-rationally oriented within this set of countries.

An interesting feature of the R^2 distribution is revealed in Figure 9. There is a negative correlation ($r=-.459$, $p<0.05$) between the national ratio of environmental action and the variance explained by all three dimensions of environmental concern. The less people are engaged in everyday action in a country, the higher the contribution of attitudes to individual action. This finding can be an indication of what Bell (2012, pp. 383–386) termed "virtual environmentalism" to describe people behaving environmentally friendly without necessarily holding environmentally friendly attitudes. Virtual environmentalism means "making environmentalism easy" (Bell 2012, p. 385) by creating smart and proper social as well as infrastructural features (see also chapter 3.2.2.1). In a similar manner, this result can also be explained in the vein of the practice theory approach (Røpke 2009, Schor 2005): In countries in which environmental action is more widely spread, it is a bigger part of the daily routine and thus does not need extra environmental concern.

Figure 9: Scatterplot: Everyday action and its variance explained by environmental concern

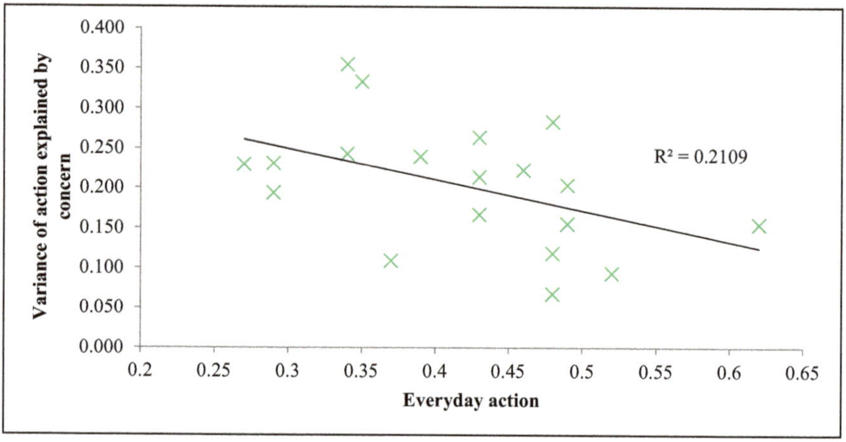

In Table 5.3 below, negative coefficients indicate a higher acceptance of environmental laws because the MCA quantified the latent variable in such a way that higher values indicate preference of voluntary environmental protection. When investigating which type of environmental concern is most important in forming the acceptance of laws, the result is clear: the evaluative dimension of environmental concern is the strongest predictor in every nation except France.[11] Second most important is the conative dimension, and the lowest effects stem from the cognitive component. This is the exact opposite of the results for everyday action,

11 The evaluative dimension is still the strongest predictor when considering the total effects (i.e. the direct and indirect effects) which are not reported.

where the cognitive dimension was the most important, followed by the conative and then the evaluative dimension.

Additionally, evaluative environmental concern is significant in each nation whereas in the other two dimensions non-significant coefficients are found. With regard to the theoretical background, preferring environmental laws over voluntary environmental protection is quite a value-rational decision expressed by the dominance of the evaluative dimension in its formation.

Despite the fact that explained variance of the acceptance of environmental laws varies to approximately the same extent as for everyday action, there is no relationship between the country-level degree of acceptance and its explained variance by environmental concern (Figure 10). When recalling the ratios of the acceptance of laws (Table 5.1) the amount that environmental concern contributes to the explanation of the ratios (Table 5.3) is rather culturally driven. For instance, in Great Britain or the USA the ratio is relatively low but the explained variance quite high. Both nations are classified a liberal welfare-state regimes where government interventions are seen more skeptically (Esping-Andersen 1990). Consequently, positive attitudes towards an issue increased the acceptance of respective laws more strongly here. In contrast, Croatia and Slovenia, which both share a communist history, hold high acceptances of environmental laws but environmental concern only contributes little to its formation. In these states the role of the state is more pronounced. Then there are countries like Sweden or Norway where both the acceptance of laws and environmental concern's contribution are high. In such "state-friendly" societies (Grendstad *et al.* 2006 see profile of Norway for further explanation) general consent to state interventions is higher but, as the numbers indicate, accepting interventions concerning the environment requires concern about the environment. Living within a state-friendly society better enables people to transfer their environmental concern into adopting the role of an environmentally friendly legal actor.

Table 5.3: Explained variances of acceptance of environmental laws by environmental concern

	R^2: Acceptance of laws	Cognitive --> Laws	Evaluative --> Laws	Conative --> Laws
Austria	0.105	0.014 (−0.133/0.180)	−0.358* (−0.493/−0.236)	−0.267* (−0.438/−0.110)
Belgium	0.171	−0.141* (−0.254/−0.022)	−0.363* (−0.493/−0.234)	−0.322* (−0.439/−0.203)
Canada	0.182	−0.146* (−0.272/−0.023)	−0.327* (−0.456/−0.228)	−0.192* (−0.282/−0.085)
Croatia	0.061	−0.182* (−0.292/−0.060)	−0.329* (−0.502/−0.181)	−0.021 (−0.115/0.078)
Czech Republic	0.084	−0.161* (−0.276/−0.057)	−0.310* (−0.449/−0.176)	−0.053 (−0.134/0.031)
Denmark	0.114	0.064 (−0.052/0.185)	−0.436* (−0.534/−0.342)	−0.105* (−0.187/−0.022)
Finland	0.190	−0.093 (−0.238/0.068)	−0.417* (−0.579/−0.282)	−0.230* (−0.352/−0.113)
France	0.137	−0.258* (−0.353/−0.164)	−0.141* (−0.233/−0.061)	−0.296* (−0.364/−0.224)
Germany-West	0.116	−0.115 (−0.250/0.030)	−0.606* (−0.850/−0.413)	−0.164* (−0.295/−0.017)
Germany-East	0.056	−0.284 (−1.058/0.135)	−0.410* (−0.826/−0.120)	0.079 (−0.110/0.306)
Japan	0.260	−0.295* (−0.414/−0.172)	−0.447* (−0.562/−0.342)	−0.420* (−0.530/−0.311)
New Zealand	0.199	−0.311* (−0.441/−0.181)	−0.432* (−0.551/−0.310)	−0.102 (−0.232/0.027)
Norway	0.264	−0.193* (−0.327/−0.051)	−0.526* (−0.693/−0.369)	−0.140* (−0.240/−0.040)
Slovenia	0.069	−0.023 (−0.151/0.106)	−0.360* (−0.482/−0.245)	−0.074 (−0.171/0.017)
Spain	0.076	−0.120* (−0.200/−0.046)	−0.232* (−0.328/−0.150)	−0.074* (−0.129/−0.020)
Sweden	0.274	0.011 (−0.118/0.168)	−0.664* (−0.842/−0.512)	−0.120* (−0.240/0.000)
Switzerland	0.046	−0.060 (−0.212/0.085)	−0.186* (−0.319/−0.074)	−0.182* (−0.301/−0.053)
Great Britain	0.253	−0.194* (−0.317/−0.078)	−0.487* (−0.622/−0.353)	−0.166* (−0.287/−0.044)
United States	0.230	−0.156* (−0.277/−0.042)	−0.555* (−0.687/−0.423)	−0.185* (−0.300/−0.077)
Total	0.139	−0.132* (−0.159/−0.102)	−0.404* (−0.434/−0.373)	−0.160* (−0.184/−0.137)

Notes: Model fit: CFI=0.939; RMSEA=0.015; SRMR=0.056; unstandardized coefficients; confidence intervals in parentheses.

Source: ISSP 2010, own calculations.

Figure 10: Scatterplot: Acceptance of environmental laws and its variance explained by environmental concern

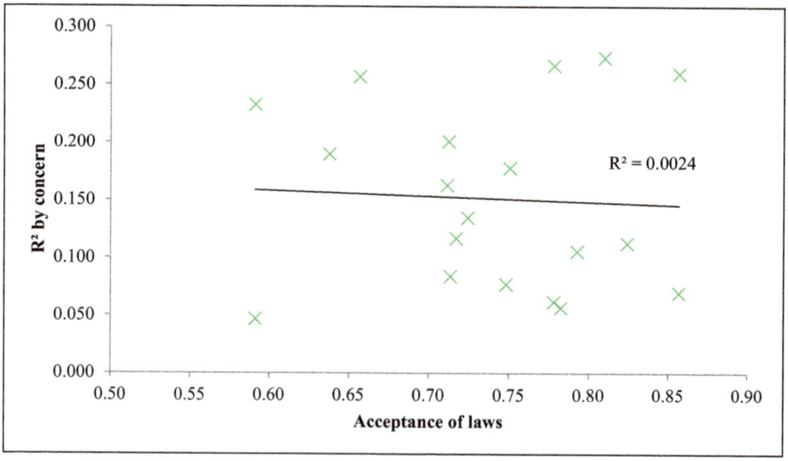

The effects of environmental concern on activism are depicted in Table 5.4. The strongest predictor is the evaluative dimension of environmental concern, followed next by the conative and least important is the perception of environmental threats. This is the same order as for the acceptance of environmental laws. Accordingly, environmental activism is more value-rationally driven than instrumental-rationally driven owning to the fact that the evaluative dimension is the most important predictor and the cognitive dimension the least important one. However, unlike for the acceptance of laws, the effect of the evaluative dimension is not significant in every nation (i.e. the Czech Republic and Japan) and also not the strongest predictor in each country (i.e. Germany East, Norway and Sweden).

Table 5.4: Explained variances of environmental activism by environmental concern

	R^2: Activism	Cognitive --> Activism	Evaluative --> Activism	Conative --> Activism
Austria	0.163	−0.023 (−0.188/0.155)	−0.560* (−0.733/−0.400)	−0.175* (−0.324/−0.007)
Flanders	0.166	−0.154* (−0.278/−0.025)	−0.450* (−0.605/−0.307)	−0.263* (−0.388/−0.127)
Canada	0.157	−0.167* (−0.296/−0.032)	−0.312* (−0.460/−0.179)	−0.285* (−0.485/−0.118)
Croatia	0.063	−0.091* (−0.155/−0.030)	−0.191* (−0.310/−0.090)	−0.088* (−0.148/−0.029)
Czech Republic	0.072	−0.148* (−0.236/−0.069)	0.011 (−0.077/0.097)	−0.254* (−0.334/−0.174)
Denmark	0.137	−0.026 (−0.187/0.141)	−0.341* (−0.495/−0.207)	−0.418* (−0.595/−0.248)
Finland	0.157	−0.133 (−0.266/0.003)	−0.315* (−0.467/−0.194)	−0.181* (−0.2889/−0.073)
France	0.132	−0.198* (−0.309/−0.081)	−0.415* (−0.553/−0.302)	−0.149* (−0.241/−0.054)
Germany-West	0.200	0.012 (−0.125/0.148)	−0.600* (−0.867/−0.386)	−0.566* (−0.784/−0.349)
Germany-East	0.091	−0.433* (−1.267/−0.064)	−0.219* (−0.464/−0.021)	−0.219* (−0.342/−0.112)
Japan	0.023	0.003 (−0.061/0.071)	−0.006 (−0.061/0.050)	−0.117* (0.167/−0.063)
New Zealand	0.207	−0.257* (−0.408/−0.105)	−0.449* (−0.619/−0.307)	−0.308* (−0.519/−0.111)
Norway	0.140	−0.125 (−0.259/0.006)	−0.186* (−0.340/−0.025)	−0.264* (−0.395/−0.142)
Slovenia	0.067	−0.063 (−0.157/0.024)	−0.227* (0.351/0.129)	−0.127* (−0.191/−0.063)
Spain	0.133	−0.065 (−0.150/0.022)	−0.434* (−0.562/−0.326)	−0.149* (−0.208/−0.090)
Sweden	0.115	−0.032 (−0.144/0.093)	−0.251* (−0.376/−0.141)	−0.284* (−0.397/−0.173)
Switzerland	0.220	−0.064 (−0.297/.0173)	−0.738* (−0.997/−0.517)	−0.362* (−0.521/−0.205)
Great Britain	0.183	−0.174* (−0.297/−0.060)	−0.352* (−0.495/−0.232)	−0.122* (−0.226/−0.024)
United States	0.137	−0.014 (−0.118/0.092)	−0.453* (−0.587/−0.341)	−0.087 (−0.190/0.007)
Total	0.126	−0.111* (−0.139/−0.083)	−0.357* (−0.389/−0.324)	−0.199* (−0.224/−0.172)

Notes: Model fit: CFI=0.937; RMSEA=0.015; SRMR=0.059; unstandardized coefficients; confidence intervals in parentheses.

Source: ISSP 2010, own calculations.

Figure 11: Scatterplot: Environmental activism and its variance explained by environmental concern

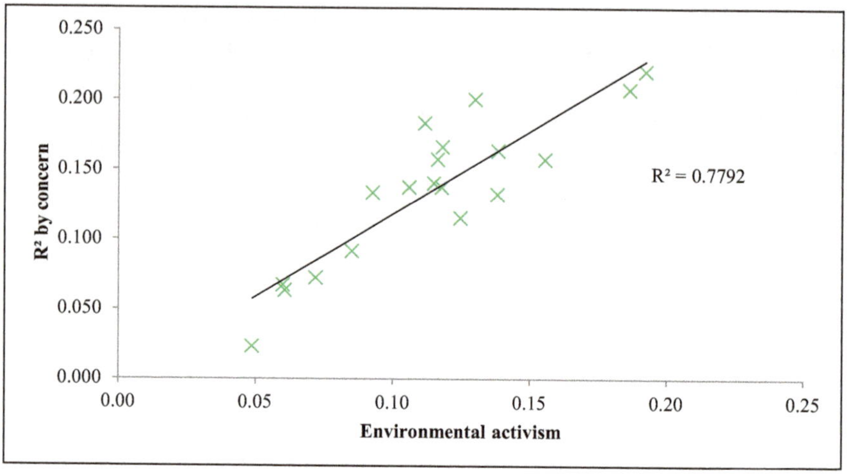

In contrast to everyday action, the greater activism exhibited by the people within a country, the higher the contribution of environmental concern (Figure 11). More environmental concern entails more activism even though the overall explained variances of environmental activism are lower than those of everyday action or the acceptance of laws. In country comparison, activism is higher where environmental concern exercises a greater impact on activism. A reason for the generally lower explained variance as well as the lower variation in size of R^2 between countries can be traced back to the low-cost hypothesis. Activism is not only often time demanding; first and foremost it is not readily available. Demonstrations do not take place every day and petitions to be signed are not in the mailbox every day. Concerning the donation of money, signing a petition or being a member of an environmental protection group not only requires availability but also involves dealing with the issue and organization. Multiple factors make environmental activism a high-cost action and therefore environmental concern does not contribute to an overly large degree.

Figure 12 depicts four types of curves showing the various relationships between an action's cost and environmental concern's contribution. The most frequent pattern is the inverted U-shape as suggested by Schan (2000). For the low-cost acceptance of laws, environmental concern is a less strong predictor than for the medium-cost everyday action. The higher cost of environmental activism goes together with a lower effect of environmental concern. Conversely, in

Belgium and Germany West the curve runs in the opposite direction. The effect of concern decreases from the acceptance of laws to everyday action and then grows again with an increased action cost. As a result of virtual environmentalism, everyday action is already widespread and needs less environmental concern. Thus the attention is shifted to more symbolic activities to express one's environmental concern: supporting laws and participating in activism. Finally, four linear trajectories are present; three negative ones (Norway, Japan and Sweden) and a positive one (Switzerland). The three negative curves are in accordance with the basic assumption of Diekmann and Preisendörfer (1998, 2003): the higher the costs, the lower the effect of concern. The line found in Switzerland runs counter to both, the inverted U-shape or the linear assumption. However, this might be the most intuitive relationship: the more environmentally concerned a person is, the higher the costs are that he or she accepts for an environmental activity. Although the most prevalent pattern is the inverted U-shape, the findings indicate that this relationship between environmental attitudes and actions is not universal but depends on country and culture. This on the one hand implies that the low-cost hypothesis and attention-shifting strategy are useful tools for this purpose but the researcher should be open to country-specific outcomes.

So far, the relationship between the three dimensions of environmental concern and the effects of concern on three types of action has been elaborated. Up to now, the question of how far the three types of action go hand in hand has been entirely neglected; or to formulate it technically, is there a correlation between them? Figure 13 depicts the correlations between the three latent variables by country. The countries are sorted in declining order of the correlation between everyday action and activism.

Figure 12: The low-cost hypothesis in cross national perspective

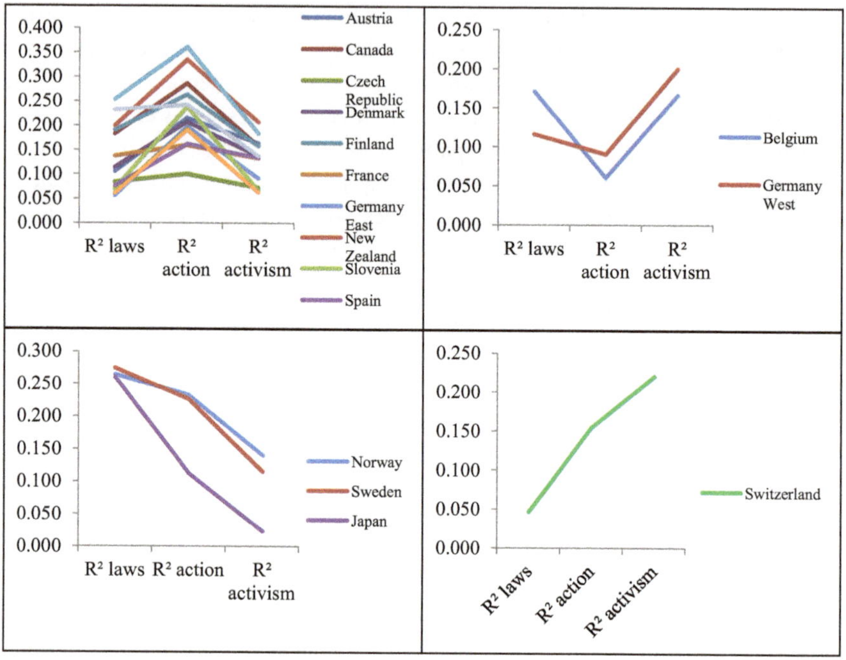

All correlations between these two variables are significant. There are pronounced country differences with a range from r=0.145 in Germany West and r=0.400 in New Zealand. It is also important to note that New Zealand, Canada and Great Britain were also among the top four concerning the amount of variance attributed to environmental concern. The correlations of the other variable combinations are lower than between everyday action and activism with the exception of Croatia where action and laws correlate a bit stronger. In Germany East, Slovenia and Switzerland the correlations between everyday action and laws are not significant. Between laws and activism this is only true for Slovenia and Germany East.

Figure 13: Correlation between the three types of environmental action

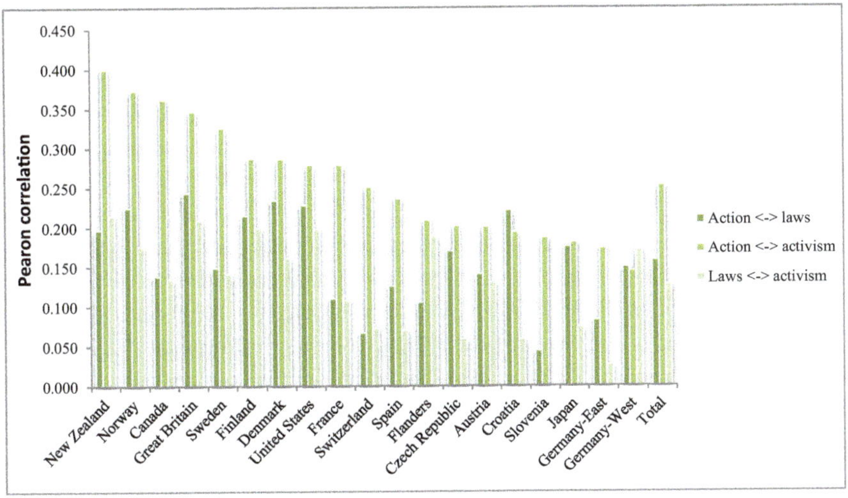

The findings just presented support the assumptions from the subjective rationality and attention-shifting strategy. A lower correlation between the variables of environmental action indicates that people choose only one action which they either rationalize as the best alternative (subjective rationality) or shift their attention to activities where the most social appraisal is expected (attention shifting). The higher the correlation, the greater the extent to which these activities go hand in hand and are less affected by subjective rationalization or attention shifting towards one specific activity.

For the micro analysis in the following chapter six countries need to be selected. The country-specific explanation why which country is chosen will be discussed in the next chapter. On the macro level a previous selection seems reasonable by clustering the nations into groups. This approach is taken by grouping them into the three welfare state types according to Esping-Andersen (1990) and more recently by Vrooman (2012):[12] the social-democratic regime, the conservative (or corporatist) regime and the liberal welfare state. As a fourth type there are nations with a socialist/communist history which are also referred to as post-socialist welfare state regimes (Kollmorgen 2009) and extend the typology to the four worlds of welfare capitalism.

12 The study shows that "generally speaking, the results confirm Esping-Andrsen's typology" (Vrooman 2012, p. 461).

The 19 countries of the ISSP used in this study can be categorized into the welfare states as follows: Finland, Denmark, Norway, and Sweden fall within the social democratic welfare regimes; the group of conservative welfare states consists of Austria, Belgium, France, Germany, Japan, Spain (the Mediterranean states are special variants of the conservative regime according to Esping-Andersen (1990) and therefore put into this group) and Switzerland; the liberal type is represented by Great Britain, Canada, New Zealand and the USA; and the post-socialist welfare states are Croatia, the Czech Republic, and Slovenia. Not all borders between the welfare state types are as clear cut as the theory suggests and some hybrid types emerged. In Vrooman (2012, p. 457) Austria, Finland, and Switzerland are described as hybrids. In the end, however, Finland is placed in the social-democratic group while Austria and Switzerland are part of the group named "corporatist/hybrid group" (Vrooman 2012, p. 469). Therefore, for the country selection in the following chapter, only the four basic types of welfare state regimes will be regarded while hybrids are ignored.

6. Country profiles

So far, the macro perspective, even though the number of countries is rather low, has unveiled differences in the amount of environmental concern and action and revealed that the 'standard' theories do not provide a linear explanation of environmentalism as it is supposed. The general micro analysis links the three dimensions of environmental concern to each other and to environmental action. There is a substantial amount of variation concerning the variance explained by environmental concern between countries, and in different nations the three dimensions play different roles. Six example country-specific profiles will provide deeper insights into the formation of environmental attitudes and action with regard to birth cohorts, gender, general values and social positions. As there are 18 countries (Germany East and West counted as one country) with sufficiently high data quality, the following criteria will serve as the guideline for choosing the six countries. As this is a cross-national comparison it is quite reasonable to choose countries from various regions of the world with differing welfare state regimes to capture the geographical and cultural diversity of the sample. As the macro analysis already showed, and as derived from the wealth and affluence hypothesis, the economic conditions of a country have an impact on people's environmentalism, therefore the six countries should also differ with regard to their economic wealth in terms of GDP per capita. Finally, as environmental attitudes and action and their relationship are the center of interest, the countries should show substantive between-country variation in the intensity of environmental attitudes and the level of action, and in the way these two aspects are linked.

The Czech Republic is not only the second-'poorest' nation in the sample (GDP per capita 2010: 18,910$) but also with a low level of variance explained, ranks second last and last in the evaluative and conative dimension respectively. The Czech Republic also has a socialist history and is referred to as the post-communist welfare state. Norway is the richest nation in the sample and is the chosen example of a social-democratic welfare state; with 22.6% of the variance explained by environmental concern it puts itself in a middle-ranking position. Besides that, the level of everyday action is lowest in Norway. Germany as a case of the conservative welfare state will also be investigated. In the Western part only 9% of the variance of everyday action can be explained while this amount is larger in the East (19.7%). Germany, in general, has the reputation of being a very environmentally friendly country (Ignatow 2006a, Franzen and Vogl 2010, Franzen *et al.* 2009) and is said to be a pioneer in green technologies. Thus it makes sense to have a closer look at this country, with a separate investigation of East and West. The USA, on the contrary, instead has the reputation of being more skeptical about environmental issues (Dietz *et al.* 2007), hence it will be investigated as well. Additionally, the USA represents a liberal welfare state regime. New Zealand, with the second-highest variance explained, will serve as the representative for Oceania while Spain seems to have a reasonable claim to be included as one of the Euro-crisis countries. New Zealand, like the USA, is a liberal welfare state and Spain as a Mediterranean nation represents a special variety of the conservative welfare regimes. The profiles will follow in alphabetic order; hence, the Czech Republic will be the first candidate.

Each of the six country profiles will start with some general prior descriptive research findings of the respective countries' environmental concern and action. These will briefly be discussed to get an insight into where the respective country has been positioned up to now in other studies. Thereafter, descriptive statistics provide an overview of the salience of the environment and environmental issues: the environment in relation to other issues, the importance of various environmental problems and the future energy supply. This recourse to prior research and the salience of the environment serves to deliver some more general information about the respective country before the more technical issue and result presentation of the multivariate analysis starts.

Tables 6.1 to 6.3 provide an average overview over the salience of the environment across the six chosen countries. These tables serve as reference, to compare country specific results with the cross-national average to help to identify outcomes that are typical for one nation but deviate from the average.

Table 6.1: Most important issues across the six countries

	Most important	Second most important	Score
The economy	3068	1385	3760.5
Health care	2201	2026	3214.0
Education	1489	1900	2439.0
Crime	566	930	1031.0
The environment	506	876	944.0
Poverty	440	666	773.0
Immigration	303	543	574.5
Terrorism	255	449	479.5
None of these	75	54	102.0

Notes: Source: ISSP 2010, own calculations.

In Table 6.1 the economy, health care, education and crime are ranked before the environment, the economy with a score 4 times and health care 3 times as high. As illustrated in the chapters 1 to 3, environmental pollution and problems can cause health problems, economy and environment are inevitably connected and education is thought of as a major predictor of environmental concern.

Besides ranking the environment in relation to other issues, the respondents also had to judge the most important environmental problem. The results are shown in Table 6.2. Averaged out across the six nations, air pollution holds first place followed by climate change. The first problem, air pollution, is a more local one reflected in the higher number of people that selected it as the most important problem for their family. The problem of climate change later clearly reflects a global, more universal problem mirrored by the much higher frequency selecting climate change as the most important problem for the country. The third-most important issue, using up natural resources, is also a more global problem. Likewise, two of the bottom three issues (nuclear waste and water shortage) represent macro (i.e. for country) issues whereas genetically modified foods pose a micro threat (i.e. for the individual and the family).

Table 6.2: Most important environmental problems across the six countries

	For country	**For family**	**Score**
Air pollution	1446	1664	1555
Climate change	1576	1074	1325
Using up our natural resources	1217	745	981
Water pollution	1018	740	879
Chemicals and pesticides	826	866	846
Domestic waste disposal	592	780	686
Genetically modified foods	397	783	590
Water shortage	655	478	567
Nuclear waste	793	308	551
None of these	103	871	487

Notes: Source: ISSP 2010, own calculations.

Table 6.3: Energy supply across the six countries

	Best source for future energy supply		Real total primary energy supply (TPES) 2010
	%	N	%
Solar, wind or water power	68.8%	5639	2.6%
Nuclear power	14.8%	1220	10.1%
Coal, oil and natural gas	6.9%	572	82.5%
Fuels made from crops	6.4%	523	4,8%
US: Fuels made from waste[a]	2.0%	162	
None of them	1.5%	120	0.0%
Total	100%	8236	100%

Notes: [a]The TPES of waste as energy supply is included in the 'fuels made from crops' category because 'fuels made from waste' are only a selectable category in the USA.

Source: ISSP 2010, own calculations. International Energy Agency (IEA) (2012)

By some distance solar, water and wind power (68.8%) are preferred over nuclear power (14.8%) or fossil fuels (6.9%). Concerning the energy supply it is thought-provoking that in terms of preference the order of the top three sources is reversed when examining their contribution to the actual TPES. This is especially remarkable because the distance between the preference of wind, solar and water power as primary energy supplier and its actual use is quite large. Across the six countries 68.8% consider these sources to be most important but only 2.6% of the primary energy is supplied in this way. Fossil fuels are still the most widespread energy sources but are, almost level with fuels made from crops, the least preferred.

To sum up, the environment is a medium-priority issue for the respondents. Air pollution is the most important and climate change the second-most important environmental problem, showing that both local and global issues are essential and there is a clear preference of renewable energies over fossil fuels on average across the six countries. This averaging quite likely veils country-specific particularities. Chapter 6.1 through 6.6 will uncover them with the Czech Republic as the first country to be put under the microscope.

6.1 The Czech Republic

GDP per capita (2010): $18,910

The communist history of the Czech Republic ended in 1989 with the transition of the former Czechoslovakia into a democracy. Finally, in 1993, Czechoslovakia was separated into the Czech Republic and the Slovak Republic (Mahoney

2010 chapter 9). As a former communist country, as Nawrotzki (2012) already found, the influence of the political position is suggested to be in reverse to that of the other countries. While a higher environmental concern is usually linked to left world views in post-communist countries, it is suggested to be the conservatives that are more environmentally concerned (see also chapter 4.2.2). This is an empirical question to which this work seeks an answer. Nawrotzki (2012, p. 295) also provides descriptive statistics of the scale of the willingness to pay for environmental protection, showing that Czech people exhibit the second-lowest willingness out of 20 countries in the ISSP 2000. Only the respondents in Bulgaria were less willing. Likewise, Kemmelmeier et al. (2002) present the same scale as Nawrotzki with the ISSP 1993 and found the Czech Republic 18th out of 22. Depending on the measure, other cross-national studies discover the Czech Republic to have rather low environmental concern (e.g. Diekmann and Franzen 1999, Franzen and Meyer 2004, 2010). Contrary to the findings just presented (in which the authors used ISSP data), Inglehart (1995) analyzed the World Values Survey (WVS) 1990/91, and found Czechoslovakia in the top ten of the most willing to pay for environmental protection out of 39. These findings fit well into the discussion about the affluence, post-materialism and globalization hypothesis (see chapter 3.2.1) where analyses with the ISSP usually find economically weaker countries to have lower environmental concern whereas in the WVS it is the other way round.[13]

Concerning private pro-environmental behavior, Hadler and Haller (2011) show that Czech people are the 8th most pro-environmentally active out of 23 countries, thus, placing themselves in an upper middle position. In a study comparing six countries, Schulz et al. (2005) find the Czech Republic to be the third most active nation after Germany and New Zealand but before Russia, Brazil and India.[14] The Czech sample considers local environmental problems to be more severe than the Germany's or New Zealand's sample, but when it comes to global environmental problems Germany is most concerned followed by the Czech Republic (Schultz *et al.* 2005). In the present study, in the frequency of everyday action, the Czech

13 One reason might be that the issue of data quality is mostly neglected in these studies (see Chapter 4.2 and Lachmann, forthcoming). Secondly, the wording of the questions used is not the same. The WVS asks for the willingness to pay *higher* prices/taxes whereas the ISSP asks for *much higher* prices/taxes.

14 The authors do not analyse a representative country sample but small samples (n=120–217) of students. They also acknowledge that therefore only characteristics of the sample, not of the country in which the sample is collected, can be compared (Schultz *et al.* 2005, p. 472).

Republic ranks 13th out of 19 and thus is in a lower middle position (Table 5.1). As seen in the Table 5.1 above, Czech people occupy last position both in the conative and evaluative dimension of environmental concern. Likewise, the relation between concern and everyday action in the Czech Republic is among lowest (Table 5.2). With regard to the perception of environmental threats, the ratio of the Czech Republic (0.65) lies somewhat above the average of 0.59.

When trying to summarize the Czech Republic's overall degree of environmentalism in comparison to other country samples, it is average to lower average.

Table CZ 1: Most important issues in the Czech Republic

	Most important	Second most important	Score
The economy	612	257	740.5
Health care	390	405	592.5
Crime	126	238	245.0
Education	80	140	150.0
The environment	67	141	137.5
Poverty	76	117	134.5
Immigration	18	49	42.5
Terrorism	29	26	42.0
None of these	12	12	18.0

Notes: Source: ISSP 2010, own calculations.

Table CZ 2: Most important environmental problems in the Czech Republic

	For country	For family	Score
Air pollution	454	389	422
Domestic waste disposal	215	189	202
Climate change	122	134	128
Using up our natural resources	174	77	126
Chemicals and pesticides	126	116	121
None of these	24	212	118
Water pollution	97	76	87
Nuclear waste	94	44	69
Water shortage	42	38	40
Genetically modified foods	31	49	40

Notes: Source: ISSP 2010, own calculations.

As Table CZ 1 illustrates, the environment clearly lies behind four other social issues: the economy, health care, crime and education. With a score of 740.5 the economy is attributed more than five times the importance as is the case with the environment. Health care is still about 4 times as important. Czech people show clear preferences for other issues. These are the same four issues that score higher in the six nation's average with the only exception that crime and education are interchanged.

When looking at Table CZ 2, two problems that have a very direct day-today effect on people are the two most important ones: air pollution and domestic waste disposal. The global phenomenon of climate change as well as using up natural resources are also relatively important. The three least central issues are the equal to the average (Table 6.2), only that nuclear waste and genetically modified foods are interchanged. Besides a small country-specific deviation, the priority ranking of various environmental problems converges with the average. To what extent this convergence is also resembled in the preference of energy sources will be shown in Table CZ 3. Support for solar, wind and water power is lowest in the Czech Republic among the countries under scrutiny as is the percentage that these energy sources contribute to the nation's energy supply (see Table 6.3 and consult the respective tables on energy supply in the other countries). However, Czech people prefer renewable energy more than the other sources. Besides that, in comparison with the other nations the support for coal, oil and natural gas greatly resembles the average while the support for nuclear power is highest in the Czech Republic. So in this respect the Czech Republic is less ecological in comparison to the other five sample countries.

Table CZ 3: Energy supply in the Czech Republic

	Best source for future energy supply		**Real total primary energy supply (TPES) 2010**
	%	N	%
Solar, wind or water power	44.9%	585	0.7%
Nuclear power	31.3%	408	16.6%
Fuels made from crops	13.7%	178	6.0%
Coal, oil and natural gas	7.2%	94	79.5%
None of them	2.9%	38	–2.9%[a]
Total	100%	1303	100%

When the export of energy exceeds the energy imports and production a negative TPES is the result.
Source: ISSP 2010, own calculations. International Energy Agency (2012).

Multivariate analysis

Base model: CFI=0.964; RMSEA=0.047; SRMR=0.047; cognitive → affective = 0.350 (0.271/0.434); cognitive → conative=0.080 (0.001/0.163); affective → conative=0.503 (0.414/0.601).

Cohorts and gender (Table CZ 4): Relatively few of the coefficients concerning cohorts and gender are statistically significant. Women are slightly more concerned about environmental issues (cognitive and evaluative) and also act in a somewhat more environmentally friendly way in their daily lives. Their higher support for laws and activism is mediated via concern. The cohort effects are more striking, not only because they are larger but in particular because they are contradictory.

Table CZ 4: Cohorts, gender and environmentalism in the Czech Republic

	Cognitive	Evaluative	Conative	Action	Action (total)	Laws	Laws (Total)	Activism	Activism (total)
Male (ref.)									
Female	0.109*	0.106*			0.127*		−0.071*		−0.034
	(0.032/0.181)	(0.014/0.193)			(0.069/0.185)		(−0.116/−0.034)		(−0.055/−0.017)
Born before 1934 (ref.)									
1934–1943									
1944–1953								−0.188*	−0.188*
								(−0.330/−0.036)	(−0.330/−0.036)
1954–1963								−0.185*	−0.185*
								(−0.323/−0.030)	(−0.323/−0.030)
1964–1973				−0.078*	−0.078*			−0.149*	−0.149*
				(−0.156/−0.001)	(−0.156/−0.001)			(−0.289/−0.006)	(−0.289/−0.006)
1974–1983				−0.143*	−0.143*			−0.235*	−0.235*
				(−0.225/−0.063)	(−0.225/−0.063)			(−0.385/−0.078)	(−0.385/−0.078)
Born after 1983			0.203*	−0.187*	−0.162*			−0.318*	−0.366*
			(0.098/0.318)	(−0.284/−0.101)	(−0.256/−0.075)			(−0.467/−0.160)	(−0.517/−0.206)
R^2	0.008	0.114	0.246	0.129	0.129	0.088	0.088	0.079	0.079
Model fit	CFI=.971; RMSEA=.037; SRMR=.034								

Notes: *p<0.05; unstandardized coefficients; bootstrapped confidence intervals below coefficient in parentheses; only significant coefficients are shown; negative coefficients on laws and activism indicate more acceptance of laws or more activism. N= 1428. Source: ISSP 2010, own calculations.

Source: ISSP 2010, own calculations.

The youngest cohort is most willing to pay but is least actively engaged in everyday action. Why is that so? The low-cost hypothesis does not provide an intuitive explanation for this non-action. When recalling that in the Czech Republic environmental concern contributes only little to the explanation of action ($R^2_{activism}$=0.072; R^2_{laws}=0.084; R^2_{action}=0.100), the Czech Republic's environmental concern in general as well as group-specific concern is rather symbolic and not very helpful in explaining the variation in action. The attention-shifting strategy might lend a helping hand. Instead of including pro-environmental action in their daily routine, the attention is shifted to other activities expressed by the higher scores of activism attained by the younger ones.

General values: Table CZ 5 shows the final model of general values. The effect of materialist values is apparent and in accordance with the post-materialism hypothesis on the micro level: materialists perceive environmental problems as less severe, are less willing to do something, and are less engaged in activism. Additionally, mixed types are also less likely activists. Those in the middle position between materialism and post-materialism do not distinguish themselves from post-materialists significantly in the other aspects. More surprising is higher support for laws as well as more frequent everyday action by both mixed types and materialists. As already mentioned in the theoretical part on post-materialism, environmental problems are very material, and action in the day-to-day routine is very material as well. Thus, at least suggestively, the higher scores in everyday action can be explained. Obviously, the materialists' world view leads likewise to a more pronounced reliance on more materialist government action: imposing laws. Another explanation would be a shifting of attention: materialists are less environmentally friendly in most respects but when there is something, the attention is shifted to a symbolic support for laws. Additional research is suggested to further elaborate why materialist values result in such contradictory outcomes.

The pattern of trust is quite obvious. It is basically the same in all 6 country samples. Higher trust entails lower cognition of environmental threats. In contrary, the evaluative component, conative environmental concern, everyday action and activism are influenced positively. Although the support for laws is not affected directly, the total effect mediated via environmental concern does increase the support. As the measure of trust in this study is a measure of generalized trust, it can be said that in the Czech Republic generalized trust can be a strong aid to overcoming environmental skeptical positions. Yet, the negative effect on the cognitive dimension is somewhat counterintuitive.

Table CZ 5: General values and environmentalism in Czech Republic

	Cognitive	Evaluative	Conative	Action (direct)	Action (total)	Laws (direct)	Laws (total)	Activism (direct)	Activism (total)
Post-materialists (ref.)									
Mixed type				0.196*	0.196*			0.240*	0.240*
				(0.082/0.321)	(0.082/0.321)			(0.066/0.411)	(0.066/0.411)
Materialists	−0.123*		−0.233*	0.257*	0.200*	−0.171*	−0.134*	0.234*	0.304*
	(−0.196/−0.049)		(−0.306/−0.147)	(0.136/0.393)	(0.076/0.332)	(−0.268/−0.067)	(−0.238/−0.033)	(0.048/0.410)	(0.121/0.480)
No answer/ preference (ref.)[a]									
Far left		−0.232*			−0.013*		0.088*		0.019*
		(−0.436/−0.055)			(−0.028/−0.003)		(0.017/0.170)		(0.003/0.042)
Left									
Central			0.291*	−0.102*				−0.252*	−0.308*
			(0.171/0.425)	(−0.202/0.002)				(−0.391/−0.110)	(−0.450/−0.165)
Right/conservative			0.168*	−0.116*	−0.095*				−0.032*
			(0.073/0.266)	(−0.201/−0.040)	(−0.178/−0.019)				(−0.058/−0.014)
Far right		−0.733*			−0.039*		0.253*		0.061*
		(−10.356/−0.065)			(−0.086/−0.005)		(0.029/0.529)		(0.008/0.133)
Non-voter		−0.136*		−0.127*	−0.134*		0.052*		0.011*
		(−0.239/−0.045)		(−0.198/−0.060)	(−0.208/−0.067)		(0.020/0.097)		(0.004/0.021)
Trust	−0.424*	0.682*	0.665*	0.182*	0.202*		−0.121*	−0.490*	−0.583*
	(−0.615/−0.238)	(0.436/0.918)	(0.447/0.877)	(0.021/0.344)	(0.043/0.378)		(−0.257/0.000)	(−0.727/−0.262)	(−0.805/−0.361)
R^2	0.023	0.151	0.302	0.133	0.133	0.098	0.098	0.095	0.095
Model fit	CFI=0.963; RMSEA=0.040; SRMR=0.033								

Notes: *p<0.05; unstandardized coefficients; bootstrapped confidence intervals below coefficient in parentheses; only significant coefficients are shown; negative coefficients on laws and activism indicate more acceptance of laws or more activism. N= 1428.

Source: ISSP 2010, own calculations.

Why are those having more trust in other people and politicians less environmentally concerned in this respect? Since this tendency is true for all nations, this issue will be discussed in the overall summary. This negative effect combined with the positive effects leads to the relatively small total effect of trust on action. Fewer of the positive effects are transferred because the strongest predictor among the three dimensions of concern is the cognitive one. The total effect of trust on laws is even weaker due to the missing direct effect while the total effect on activism is largest resulting from the relatively large direct coefficient.

The most intriguing pattern is revealed concerning the political position. Only the far left, the far right as well as the non-voters differentiate from those who did not give an answer to the question about their political orientation with regard to the evaluative dimension. Non-voters, far-right and far-left voters are more skeptical about the environment than the rest. In contrast, the willingness to pay is higher among central and right/conservative voters. Why so? This outcome, at least in part, supports the findings of Nawortzki (2012) that in former communist states conservatives are more environmentally concerned. When everyday action is examined, the effect, however, is negative (direct and total). The negative effect of the non-voters entails the argument that those who do not vote, for instance, because they are more skeptical about politics, are also more cynical about other issues such as the environment. This is supported by the negative effect on everyday action, less support for laws and less activism. Concerning environmental laws only the far right and far left exhibit a significant total effect indicating lower acceptance. With regard to the political position there are only the central voters that more often participate in political environmental action directly. Indirectly the right/conservatives are also more likely activists compared to the far left which records marginally higher scores (i.e. lower activism) than the far right.

Social positions (Table CZ 6): Because in the community size variable there are only four respondents that gave no answer, these are included on the reference category. The five that provided no answer about their education are treated similarly. Concerning education in the Czech Republic the reference category was set to 'lowest formal education' because there are only two persons in the data set that state they had no formal education at all. These two are included in the reference category.

Considerable effects regarding the size of the coefficients (Table CZ 6) emerge from income, education, the person's occupational status as well as from the community size.

Table CZ 6: Social positions and environmentalism in Czech Republic

	Cognitive	Evaluative	Conative	Action	Action (total)	Laws	Laws (total)	Activism	Activism (total)
Equivalence income			0.125*		0.016*				−0.027*
			(0.083/0.168)		(0.009/0.025)				(−0.041/−0.016)
Equivalence income²								−0.019*	−0.019*
								(0.036/0.000)	(0.036/0.000)
Lowest formal education (ref.)									
Intermediate						−0.184*	−0.184*		
						(−0.353/−0.013)	(−0.353/−0.013)		
Secondary		0.202*	0.110*		0.032*	−0.170*	−0.241*	−0.184*	−0.238*
		(0.106/0.289)	(0.026/0.199)		(0.018/0.050)	(−0.347/−0.002)	(−0.417/−0.075	(−0.275/−0.097)	(−0.330/−0.149)
University incomplete	0.402*				0.091*	−0.385*	−0.487*	−0.391*	−0.467*
	(0.110/0.680)				(0.027/0.167)	(−0.790/0.00)	(−0.888/−0.066)	(−0.720/−0.054)	(−0.792/−0.121)
University complete		0.345*	0.292*		0.069*	−0.257*	−0.379*	−0.217*	−0.335*
		(0.191/0.506)	(0.157/0.447)		(0.042/0.109)	(−0.504/−0.046)	(−0.616/−0.167)	(−0.363/−0.057)	(−0.489/−0.187)
Never had paid work (ref.)									
No paid work			−0.276*	0.202*	0.168*			0.282*	0.340*
			(−0.423/−0.131)	(0.092/0.314)	(0.057/0.281)			(0.125/0.432)	(0.190/0.490)
In paid work			−0.191*	0.130*	0.116*			0.238*	0.262*
			(−0.332/−0.060)	(0.025/0.238)	(0.012/0.229)			(0.097)	(0.123/0.410)
No answer				0.342*	0.342*	0.655*	0.655*		
				(0.020/0.676)	(0.020/0.676)	(0.087/10.157)	(0.087/10.157)		
School children in household				−0.063*	−0.067*			0.096*	0.103*
				(−0.111/−0.202)	(−0.115/−0.024)			(0.028/0.166)	(0.034/0.172)

Table CZ 6: Continued

	Cognitive	Evaluative	Conative	Action	Action (total)	Laws	Laws (total)	Activism	Activism (total)
Toddlers in household					−0.004*				0.006*
					(−0.007/−0.002)				(0.003/0.012)
Big city (ref.)									
Suburbs		−0.242*	0.193*				0.091*		
		(−0.440/−0.052)	(−0.004/0.367)				(0.021/0.178)		
Town			0.103*		0.013*			0.142*	0.105*
			(0.009/0.189)		(0.002/0.028)			(0.048/0.241)	(0.013/0.221)
Village			0.191*		0.024*	0.116*	0.116*	0.117*	
			(0.085/0.293)		(0.010/0.043)	(0.000/0.227)	(0.000/0.227)	(0.013/0.234)	
Farm						1.308*	1.308*		
						(.577/2.037)	(.577/2.037)		
R^2	0.022	0.144	0.302	0.122	0.122	0.104	0.104	0.106	0.106
Model fit		CFI=.972; RMSEA=.032; SRMR=.029							

Notes: *p<0.05; unstandardized coefficients; bootstrapped confidence intervals below coefficient in parentheses; only significant coefficients are shown; negative coefficients on laws and activism indicate more acceptance of laws or more activism. N= 1428.

Source: ISSP 2010, own calculations.

As proposed by the prosperity hypothesis, a higher income leads to a higher degree of willingness to pay for environmental protection. This minimally transfers into a higher degree of everyday action and activism. Even though the main effect of income on activism is not significant, the effect of the squared income is. Despite the zero-effect of income in total, from one tipping point on activism slightly decreases with an increasing income. No other non-linearity effects are present.

Respondents holding a university degree are also more willing to pay, independent of their income, and they evaluate environmental issues in a more environmentally friendly way. The latter is also the case for individuals with secondary education. People with an incomplete university degree, most likely students, perceive environmental issues as somewhat more dangerous. In general those with a better education are more environmentally concerned at least in one dimension than those with low or intermediate education. The educational groups do not differ in the amount of everyday action directly. The total effects of education that are significant lack practical relevance due to their small size. The effects of the cognitive and conative dimension on action are not strong enough (see Table 5.2) to mediate educational effects. Likewise, strongest educational coefficients are on the evaluative dimension which does not affect everyday action at all. With regard to the acceptance of laws and activism, education increases both acceptance and activism, somewhat more so for respondents holding an incomplete university degree than for those with a complete one.

Most interesting are the results of the occupational position. Those who are currently not employed as well as those with an occupation are less willing to pay than people that so far never had a paid work. The same two groups are more engaged in pro-environmental activities but record lower scores for activism. The negative effects on the willingness to pay are close to the absolute value of those on action. Due to the negative sign they diverge significantly. Although in general the willingness to do something increases everyday action (b=0.118, p=.001), in this case a reversed causality might yield an explanation. Because these people are already more engaged in these activities they are less willing to do more. The explanation for less activism is straightforward: those currently in employment have less time available for activism. People who are not in employment are mostly older people many of which are already retired. As seen in Table CZ 4 the older respondents are more active. The outcome of this correlation with age is alluded to in the full model below.

The results of the family situation are more straightforward. The number of children or toddlers in the household does not influence people's environmental concern. Looking at the total effects, in line with the hypothesis, people with children or toddlers are less active in their everyday routine and activism. This is in part due to the direct effects and in part mediated via household income (which is lower the more children/toddlers a household has). However, these are just minimal deviations from the average action/activism and lack substantive relevance.

The size of the community has some influences on environmental concern and a practically relevant impact on action. People living in the suburbs are less concerned in the evaluative dimension but more willing to pay. More willing to pay are also people living in towns or villages. The acceptance of laws is much lower among people from farms in the countryside and a little lower in villages. Activism is decreased when living in a town or village but when taking the mediation via the positive effects on the conative dimension into account, only people from towns are less active. Community size seems to be a proxy for community-based incentives or disincentives that are in competition to one another either causing a positive sign in the one aspect and a negative one in another or leading to zero coefficients. With simple dummy variables capturing the community size, there is too much information that is not gathered. All aspects that are linked to community size but not measured as a distinct variable are lumped together in one and thus lead to the contradictory or non-significant effects.

Full model (Table CZ 7): In the full model little changed regarding the size and direction of the coefficients. Some became somewhat smaller, other increased a little. The first notable changes appear among the materialists under control of the birth cohorts and social positions. Materialists and mixed types are no longer less often active than post-materialists, leaving just a small (practically negligible) significant total effect. The effects of the political position on the evaluative dimension disappear once controlling for birth cohorts. Especially, the drop of the far-right from b=–.733 in Table CZ 5 to being no longer significantly different from zero is remarkable. In the same manner, to a lower extent though, the non-voters and far-left coefficient lost significance. This in turn also leads to the discontinuation of the total effects on the acceptance of laws.

Once the dummies for the birth cohorts and gender are included in the model, the ambiguous coefficients of the occupational position (compare Table CZ 6) disappear for two reasons. First, older people register a higher amount of action. Second, the largest number of people with no paid job is older and already retired.

Table CZ 7: Full model Czech Republic

	Cognitive	Evaluative	Conative	Action	Action (total)	Laws	Laws (total)	Activism	Activism (total)
Gender	0.118*	0.101*		0.081*	0.112*		−0.072*		−0.029*
	(0.044/0.201)	(0.019/0.184)		(0.022/0.146)	(0.050/0.180)		(−0.110/−0.038)		(−0.051/−0.012)
Born before 1934 (ref.)									
1934–1943									
1944–1953					0.002*			−0.208*	−0.210*
					(0.000/0.004)			(−0.335/−0.088)	(−0.341/−0.092)
1954–1963								−0.265*	−0.265*
								(−0.413/−0.126)	(−0.413/−0.126)
1964–1973								−0.292*	−0.292*
								(−0.434/−0.138)	(−0.434/−0.138)
1974–1983				−0.087*	−0.087*			−0.383*	−0.383*
				(−0.167/−0.008)	(−0.167/−0.008)			(−0.557/−0.231)	(−0.557/−0.231)
Born after 1983			0.118*	−0.145*	−0.130*			−0.299*	−0.320*
			(0.003/0.222)	(−0.227/−0.053)	(−0.212/−0.042)			(−0.465/−0.144)	(−0.496/−165)
Post-materialists (ref.)									
Middle				0.165*	0.165*				
				(0.063/0.282)	(0.063/0.282)				
Materialists	−0.125*	−0.096*	−0.180*	0.216*	0.160*	−0.191*	−0.120*		0.065*
	(−0.204/−0.048)	(−0.202/−0.007)	(−0.255/−0.105)	(0.101/0.343)	(0.047/0.286)	(−0.285/−0.088)	(−0.219/−0.021)		(0.040/0.097)
No Answer (ref.)									
Far Left									

Table CZ 7: Continued

	Cognitive	Evaluative	Conative	Action	Action (total)	Laws	Laws (total)	Activism	Activism (total)
Left									
Central			0.253*		0.031*			−0.242*	−0.288*
			(0.124/0.390)		(0.015/0.057)			(−0.433/−0.069)	(−0.486/−0.115)
Right/Conservative			0.121*	−0.082*					−0.025*
			(0.023/0.229)	(−0.154/−0.006)					(−0.051/−0.008)
Far Right									
Non-voter				−0.066*	−0.066*				
				(−0.133/0.001)	(−0.133/0.001)				
Trust	−0.437*	0.592*	0.588*	0.175*	0.179*			−0.440*	−0.507*
	(−0.638/−0.232)	(0.350/0.909)	(0.366/0.805)	(0.025/0.351)	(0.017/0.345)			(−0.722/−0.208)	(−0.801/−0.284)
Equivalence income			0.116*		0.014*				
			(0.079/0.159)		(0.008/0.023)				
Equivalence income²								−0.022*	−0.022*
								(−0.046/−0.005)	(−0.046/−0.005)
Lowest education (ref.)									
Intermediate						−0.213*	−0.213*		
						(−0.388/−0.018)	(−0.388/−0.018)		
Higher secondary		0.179*			0.015*	−0.198*	−0.264*	−0.156*	−0.177*
		(0.095/0.270)			(0.008/0.026)	(−0.382/−0.003)	(−0.440/−0.068)	(−0.248/−0.069)	(−0.277/−0.093)
University incomplete	0.390*				0.091*	−0.414*	−0.521*	−0.354*	−0.433*
	(0.170/0.654)				(0.043/0.172)	(−0.725/0.000)	(−0.838/−0.112)	(−0.983/0.000)	(−10.056/0.000)

Table CZ 7: Continued

	Cognitive	Evaluative	Conative	Action	Action (total)	Laws	Laws (total)	Activism	Activism (total)
University		0.296*	0.165*		0.046*	−0.302*	−0.441*	−0.182*	−0.249*
		(0.164/0.422)	(0.013/0.295)		(0.023/0.076)	(−0.534/−0.064)	(−0.634/−0.174)	(−0.368/−0.012)	(−0.441/−0.086)
Never had paid work (ref.)									
No paid work									
In paid work					0.009*			0.114*	0.101*
					(0.005/0.015)			(0.005/0.228)	(0.000/0.214)
NA work									
School children in household				−0.071*	−0.074*			0.118*	0.110*
				(−0.121/−0.025)	(−0.124/−0.027)			(0.008/0.212)	(0.050/0.176)
Toddlers in household					−0.003*			0.106*	0.122*
					(−0.006/−0.001)			(0.047/0.169)	(0.012/0.217)
Big city (ref.)									
Suburbs		−0.192*			−0.010*		0.076*		0.014*
		(−0.398/0.000)			(−0.024/0.000)		(0.000/0.166)		(0.000/0.033)
Town								0.154*	0.154*
								(0.061/0.256)	(0.061/0.256)
Village			0.095*		0.012*	0.127*	0.127*	0.139*	0.122*
			(0.005/0.172)		(0.002/0.024)	(0.014/0.242)	(0.014/0.242)	(0.039/0.241)	(0.019/0.225)
Farm						1.259*	1.259*		
						(.027/1.973)	(.027/1.973)		

Table CZ 7: Continued

	Cognitive	Evaluative	Conative	Action	Action (total)	Laws	Laws (total)	Activism	Activism (total)
Cognitive		0.354*	0.122*	0.203*	0.236*	−0.151*	−0.278*	−0.115*	−0.202*
		(0.266/0.445)	(0.038/0.209)	(0.144/0.273)	(0.174/0.306)	(−0.267/−0.057)	(−0.385/−0.191)	(−0.236/−0.081)	(−0.289/−0.127)
Evaluative			0.417*		0.051*	−0.369*	−0.369*		−0.076*
			(0.328/0.530)		(0.029/0.080)	(−0.490/−0.245)	(−0.490/−0.245)		(−0.177/−0.046)
Conative				0.121*	0.121*			−0.180*	−0.180*
				(0.071/0.172)	(0.071/0.172)			(−0.262/−0.113)	(−0.262/−0.113)
R^2	0.039	0.173	0.334	0.155	0.155	0.111	0.111	0.129	0.129
Model Fit		CFI=0.981; RMSEA=0.027; SRMR=0.022							

Notes: *p<0.05; unstandardized coefficients; bootstrapped confidence intervals below coefficient in parentheses; only significant coefficients are shown; negative coefficients on laws and activism indicate more acceptance of laws or more activism. N= 1428.

Source: ISSP 2010, own calculations.

Consequently, the effect of having no paid job is a selection effect. Those who have no job due to retirement selected in to this category and in that way created the significant coefficient. However, people in paid work still post lower scores in activism which is in accordance with the assumption that when having a job, less time is available for such activities.

The effect of the number of children and toddlers in the household on activism increases in size when controlling for values and cohorts/gender. Especially the relationship to the cohorts (the 1974–83 and 1964–73 groups have more children in their household) seems to influence the impact on activism: when controlling for the presence of children/toddlers the increase in activism occurs most likely due to the time constraint imposed by children as proposed by the biographic availability assumption.

To summarize, all three aspects (i.e. cohorts/gender, values and social positions) in some way contribute to the explanation of environmentalism. Taken together, by comparing the explained variances to the basic models in tables 5.2–5.4 the overall contribution is not very large. This is so in spite of the fact that in the Czech Republic environmentalism is not yet that widespread. More personal factors, such as individual experiences or histories (Brechin 1999, p. 807) may contribute more to the explanation of environmentalism in the Czech Republic than those covered here. The changes in the R^2 by the inclusion of additional factors will be alluded to in the general summary after the country profiles. Now, moving further westwards, the Czech Republic's neighbor Germany is the next candidate under investigation.

6.2 Germany

GPD per capita (2010): $40,164

Germany is said to be a very 'environmental country' and German people are considered relatively environmentally friendly (e.g. Ignatow 2006b, Franzen and Vogl 2010, Franzen *et al.* 2009). Studies have shown that environmental concern in fact is high but this is actually not true in international comparison where Germany only occupies a middle position (Franzen and Meyer 2004, 2010, Franzen and Vogl 2010, Franzen *et al.* 2009, Kuckartz 1997). The authors attest to a relatively stable environmental concern in the 1990s and 2000s. Concerning production and development of renewable energy technologies, Germany does still occupy a leading position, "with one of every two wind power generators and one of every three solar panels in the world coming from Germany" (Zitzer 2009, p. 1). To go one step further, on July 31, 2011 federal president Christian Wulff executed the law to change the Atomic Energy Act and thus settle the energy transition or

'Energiewende' away from nuclear power toward renewable energy sources (Bundespräsidialamt 2011).

As Germany is a special case because it was previously divided into the Federal Republic (FRG) and the German Democratic Republic (GDR) and unified only on October 3 in 1990, Preisendörfer (1996b) explored the differences of environmental attitudes and behavior between Eastern (former GDR) and Western Germany and finds an "Ökologie Defizit" (ecology deficit) in the eastern part (Preisendörfer 1996b, p. 1). This means that people in Germany East are less concerned about the environment and act in a less environmentally friendly way than in the West. The author used the data of the ISSP 1993. The analysis was based on the idea of the dominance of the economy over ecology after the reunion (Preisendörfer 1996b, p. 2). The former GDR discovered economic benefits of capitalism, albeit being 'blind' to problems that go with the benefits. The ecology deficit implies that Eastern and Western Germany should be investigated separately.

In the tables depicting the salience of the environment in Germany (GER 1–3) the results for Germany West and East are presented next to each other in one table giving them an appearance different from the other countries' tables. Even though there are minor discrepancies between the two regions, maybe just the result of sampling, all in all the tables' outcomes look quite similar.

In Table GER 1 the environment ranks fourth in the West and fifth in the East and for poverty it is the other way round. Thus the ecology deficit of the East Germans, a minor one though, is represented in these findings. In the top three, education and economy changed their position which may be an indication for the still existing economic differences between both parts. Altogether Germany approximates the six-country average, with the environment in a middle-ranking position.

Table GER 1: Most important issues in Germany West and East

	Germany West				Germany East		
	Most important	Second most important	Score		Most important	Second most important	Score
Education	268	208	372	The economy	115	72	151
Health care	203	227	316.5	Health care	103	92	149
The economy	238	149	312.5	Education	90	93	136.5
The environment	67	120	127	Poverty	32	40	52
Poverty	74	88	118	The environment	20	37	38.5
Crime	36	71	71.5	Crime	16	36	34
Immigration	27	43	48.5	Immigration	14	14	21
Terrorism	23	24	35	Terrorism	4	7	7,5
None of these	6	6	9	None of these	1	0	1

Notes: Source: ISSP 2010, own calculations.

With regard to the most important environmental issue (Table GER 2) Germany West and East are also quite equal. Climate change is the most important issue for both regions, nuclear waste the second one in the West followed by air pollution, while in the former GDR states it is the other way round. The high ranking of nuclear waste underlines Germany's particular objection to nuclear power in comparison to other countries. On average, nuclear waste is the last substantive environmental issue, only followed by the category 'none of these' (Table 6.2).

When looking at energy supply preferences (Table GER 3) a slight discrepancy toward nuclear power between East and West is noticeable too. In the West only 4.9% consider this energy source the most important, and thus it is ranked last. Eastern Germans give somewhat more priority to nuclear power (8.1%) and prioritize it over 'fuels made from crops'. Nevertheless, 8.1% is still the third lowest number among the six countries, only fallen short of by Norway (5.9%) and Germany West (4.9%). 81.0% and 76.4% people in Germany West and East respectively prefer solar, wind and water power, exceeding the average of 68.5%. In comparison to New Zealand (32.5%) or Norway (31.4%), a 2.1% total primary energy supply (TPES) in Germany is quite low whereas the use of fossil fuels (78.1%) is almost as high as in the Czech Republic (79.5%) or Spain (75.9%). For comparison, Norway and New Zealand get 61.3% and 60.8% respectively of their energy from fossil fuels and only the USA (84.1%) exceeds the 80%-mark. Despite the energy transition, Germany still has to catch up with other countries.

Table GER 2: Most important environmental problems in Germany West and East

	Germany West				Germany East		
	For country	For family	Score		For country	For family	Score
Climate change	255	181	218	Climate change	97	72	85
Nuclear waste	241	62	152	Air pollution	57	60	59
Air pollution	132	132	132	Nuclear waste	96	19	58
Using up our natural resources	97	93	95	Using up our natural resources	43	41	42
Genetically modified foods	59	117	88	Genetically modified foods	29	49	39
None of these	16	125	71	None of these	3	58	31
Chemicals and pesticides	47	80	64	Chemicals and pesticides	25	33	29
Water pollution	40	26	33	Domestic waste disposal	17	20	19
Domestic waste disposal	17	44	31	Water pollution	20	13	17
Water shortage	22	12	17	Water shortage	10	8	9

Notes: Source: ISSP 2010, own calculations.

Table GER 3: Energy supply in Germany West and East

Germany West			Germany East			GER total
	Best source for future energy supply West			Best source for energy supply East		Real total primary energy supply (TPES) 2010
	%	N		%	N	%
Solar, wind or water power	81.0%	731	Solar, wind or water power	76.4%	301	2.1%
Coal, oil and natural gas	7.2%	65	Coal, oil and natural gas	10.2%	40	78.1%
Fuels made from crops	5.3%	48	Fuels made from crops	3.3%	13	8.9%
Nuclear power	4.9%	44	Nuclear power	8.1%	32	11.2%
None of these	1.7%	15	None of these	2.0%	8	-0.4%
Total	100,0%	903	Total	100.0%	394	100%

Notes: Source: ISSP 2010, own calculations. International Energy Agency (2012).

As seen in the descriptive analysis in chapter 5 (see Table 5.1) Germany East ranks lower than Germany West in the cognitive and conative dimensions of environmental concern as well as environmental action but higher in the evaluative dimension. The acceptance of laws is higher in Germany East while activism is lower. The differences between East and West in the cognitive and conative dimension, acceptance of laws and activism are statistically significant. Those of the evaluative dimension and everyday action are not.[15] In 2010 the ecology deficit still exists in some respects but the gap seems to be closing. Germany West, in the present study, is among the most environmentally concerned nations. They rank second in the cognitive dimension, occupy a middle position in the evaluative dimension, third in the conative dimension, and they exhibit the second highest amount of environmental everyday action after France. Regarding the acceptance of laws, Western Germany can be found in a lower middle position while activism is 6th highest in the West (Table 5.1). The variance of everyday action explained by environmental concern is second lowest.

Tables 5.2–5.4 show quite distinct relationships between attitudes and action in the two parts of Germany adding another justification to keeping Eastern and Western Germany separated in the multivariate analysis. For instance, in Germany West only the evaluative dimension contributes moderately to everyday action. In Germany East, in contrast, the cognitive and conative components have an impact and not the evaluative dimension. Interestingly, the dimensions

15 East-West differences: Mann-Whitney-U-tests: cognitive: $Z=-3.720$, $p=0.001$; evaluative: $Z=-0.250$, $p<0.803$; conative: $Z=-2.643$, $p<.008$; action: $Z=-1.379$, $p=.168$; laws: $Z=-3.054$, $p=0.002$; activism: $Z = -3.570$, $p=0.000$.

that contribute to the explanation of action in one part are those in which this part holds the lower values. In Germany West, environmental action is rather value-rationally driven while in Germany East it is more instrumentally rational. Additionally, environmental concern in Germany East (R^2=0.203) is much more important than in Germany West (R^2=0.093) for the explanation of everyday action. In Germany West virtual environmentalism is more prevalent. In the East, environmental action still depends more strongly on environmental attitudes. For the acceptance of laws and activism it is the other way round. Attitudes contribute more to their explanation in Germany East then in Germany West (compare Table 5.3 and 5.4).

Multivariate analysis

Germany East: Cognitive → affective =n.s.; cognitive → conative=n.s.; affective → conative=0.511 (0.325/0.743).

A look at the basic model of Germany East reveals some degree of misspecification (CFI=0.913; RMSEA=0.62; SRMR=0.057) which implies an acceptable rather than a good fit. The misspecification is not very large but would need further assessment to find the factor causing this misfit before proceeding with the analysis. Investigating the factor loading shows that Q20a (pollution by cars) and Q20e (global warming) have relatively low factor loadings (f=0.320 and f=0.246, respectively). This entails a fairly strong implied parameter for the error correlation between these two items suggested by the modification indices. Deleting Q20e does improve the fit but does not change the relations of the cognitive dimension to the other dimensions as does allowing the covariance between their residuals to be estimated freely.[16] Deleting both variables does again improve the fit but does not change the substantive results. Hence, for comparability reasons the basic model is retained as it is because the fit is at least acceptable.

The model fit of the basic model is unproblematic in Germany West: CFI=0.956; RMSEA=0.046; SRMR=0.047. Cognitive → affective=0.171 (0.096/0.249); cognitive → conative=0.115 (0.028/0.181); affective → conative=0.353 (0.211/0.493).

Cohorts and gender (Tables GER 4 and 5): Contrary to the general assumption, in Germany East women are less environmentally concerned in the evaluative dimension than men (Table GER 4). In addition women are more reluctant to accept laws. Everyday action and activism are only affected indirectly. More substantial are the cohort effects. People born in the mid-sixties to mid-seventies

16 It would be reasonable to suggest that in Eastern Germany pollution by cars are perceived as one major driver of global warming and thus these two form their own factor.

show more evaluative concern than the older or younger generations which entails also more action, support for laws and activism. This cohort grew up in the time when the environment became an important topic (Huber 2011: 79). For later cohorts, in contrast, it looks quite different. The two youngest cohorts are found to show less action and the youngest also support laws less strongly. The 1974–83 cohort also is less engaged in activism. This is quite similar tendency to that already discovered in the Czech Republic with slight differences: unlike in the Czech Republic the younger cohorts do not show higher environmental concern and are not more often engaged in environmental activism than in Germany East. Despite these effects just described, in total in Germany East the birth cohorts and gender do not contribute much to the explanation of environmentalism.

The results for Germany West (able GER 5) differ from those in the East. Women only deviate from men in the cognitive (positively) and the conative (negatively) dimension, the two aspects which do not affect everyday action in the West. In contrast, in Germany East only the evaluative dimension is negatively influenced by a gender effect, again the aspect without any impact on everyday action. The total effect on everyday action is significant but at the same time microscopically small.

Concerning the cohort effects, there is a tendency towards an inverted U-shaped curve in the evaluative dimension. Least concerned are those that were born before 1934. Until the 1954–1963 cohort there is a constant increase in the effect sizes. From 1964–73 onwards the effects decrease again. The two most concerned generations are those born between 1954 and 1973. Mediated via environmental concern, the same tendency also holds for the acceptance of laws and activism.

Table GER 4: Cohorts, gender and environmentalism in Germany East

	Cognitive	Evaluative	Conative	Action	Action (total)	Laws	Laws (total)	Activism	Activism (total)
Male (ref.)									
Female		–0.216*			–0.029*	0.175*	0.243*		0.074*
		(–0.734/–0.068)			(–0.068/–0.009)	(–0.001/0.364)	(0.064/0.429)		(.021/.163)
Born before 1934 (ref.)									
1934–1943									
1944–1953									
1954–1963									
1964–1973		0.373*			0.050*		–0.111*		–0.128*
		(0.663/0.137)			(0.017/0.104)		(–0.231/–0.035)		(–0.275/–0.045)
1974–1983				–0.560*	–0.560*			0.299*	0.299*
				(–0.789/–0.327)	(–0.789/–0.327)			(0.059/0.541)	(0.059/0.541)
Born after 1983				–0.482*	–0.482*	0.671*	0.671*		
				(–0.716/–0.256)	(–0.716/–0.256)	(0.404/0.945)	(0.404/0.945)		
R^2	0.000	0.071	0.197	0.304	0.304	0.102	0.102	0.094*	0.094
Model fit	CFI=.940; RMSEA=.044; SRMR=.049								

Notes: *p<0.05; unstandardized coefficients; bootstrapped confidence intervals below coefficient in parentheses; only significant coefficients are shown; negative coefficients on laws and activism indicate more acceptance of laws or more activism. N=418.

Source: ISSP 2010, own calculations.

Table GER 5: Cohorts, gender and environmentalism in Germany West

	Cognitive	Evaluative	Conative	Action	Action (total)	Laws	Laws (total)	Activism	Activism (total)
Male (ref.)									
Female	0.104*		−0.108*		0.015*				
	(0.019/0.193)		(−0.187/−0.041)		(0.003/0.035)				
Born before 1934 (ref.)									
1934–1943		0.228*		0.227*	0.315*		−0.148*		−0.175*
		(0.064/0.405)		(0.102/0.356)	(0.175/0.467)		(−0.273/−0.033)		(−0.322/−0.039)
1944–1953	−0.127*	0.365*		0.156*	0.277*		−0.205*		−0.261*
	(−0.251/−0.007)	(0.204/0.553)		(0.022/0.292)	(0.134/0.445)		(−0.356/−0.091)		(−0.427/−0.130)
1954–1963		0.488*			0.187*		−0.319*	−0.247*	−0.630*
		(0.319/0.657)			(0.111/0.284)		(−0.474/−0.203)	(−0.420/−0.092)	(−0.861/−0.444)
1964–1973		0.438*			0.168*		−0.288*		−0.347*
		(0.282/0.623)			(0.099/0.262)		(−0.434/−0.172)		(−0.508/−0.214)
1974–1983	0.162*	0.371*			0.165*		−0.291*		−0.334*
	(0.035/0.288)	(0.204/0.553)			(0.096/0.258)		(−0.435/−0.166)		(−0.501/−0.196)
Born after 1983		0.255*		−0.265*	−0.167*	0.385*	0.219*		−0.201*
		(0.086/0.441)		(−0.407/−0.120)	(−0.320/0.001)	(0.190/0.579)	(−0.001/0.444)		(−0.427/−0.062)
R^2	0.023	0.124	0.147	0.142	0.142	0.132	0.132	0.207	0.207
Model fit		CFI=.965; RMSEA=.036; SRMR=.035							

Notes: *p<0.05; unstandardized coefficients; bootstrapped confidence intervals below coefficient in parentheses; only significant coefficients are shown; negative coefficients on laws and activism indicate more acceptance of laws or more activism. N=989.

Source: ISSP 2010, own calculations.

With regard to everyday action, there is a drift showing that the World War II and post-war cohort is more engaged than the rest. The youngest in the sample are the least active among all cohorts. The stronger activity of the war generation may be explained by these people being used to live with scarcity of resources and thus using resources such as energy or water more sensibly. This leads to a nearly linear decline of the total effects on action from the 1934–43 cohort to those born between 1974 and 1983 with a larger downward drop for the youngest respondents. The youngest generation is also less willing to accept laws. All remaining effects on the acceptance of laws as well as on activism are only indirect effects mediated via environmental concern. In comparison to Germany East, the cohort effects are more substantive and tell us more about the variation in environmental concern.

General values (tables GER 6 and GER 7): The results including the general values measures in Germany East are somewhat unsatisfactory. Both mixed types and materialists record lower scores in the evaluative dimension than post-materialists. This also leads to significant total effects on everyday action, the acceptance of laws as well as activism of materialists and mixed types. As in the other countries, more trustful people perceive environmental threats as less dangerous but are more willing to pay. Trust is mediated only to laws and activism. Regarding the political position, only the right/conservatives (cognitive) and non-voters (evaluative) are less concerned in the cognitive and evaluative dimension. The far left and right/conservative hold the highest values in action. The total effect on everyday action of the right/conservatives is insignificant due to the negative path directed at the cognitive dimension. Support for laws as well as activism is higher among the left and far left. In contrast to the lower concern, non-voters also support laws more strongly. The results for general values in Germany East provide some clue that environmental concern and action diffused into society concerning people's general values, a hint at 'virtual environmentalism'. However, this process is far from being finished as the effects, that were found, show. The influence of general values on environmental concern in Germany West is more clearly pronounced than in the East. Materialists and mixed types are less environmentally concerned with regard to the evaluative component than post-materialists. Concerning the cognitive dimension, the coefficients are also negative but not significant. The negative effects transfer into lower scores in everyday action, the acceptance of laws and activism. In addition, activism is less wide-spread among mixed types and materialists.

Table GER 6: General values and environmentalism in Germany East

	Cognitive	Evaluative	Conative	Action	Action (total)	Laws	Laws (total)	Activism	Activism (total)
Post-materialists (ref.)									
Mixed type		−0.240*			−0.029*		0.079*		0.086*
		(−0.433/−0.075)			(−0.069/−0.008)		(0.016/0.184)		(0.017/0.198)
Materialists		−0.287*			−0.035*		0.094*		0.100*
		(−0.544/−0.021)			(−0.096/−0.004)		(0.013/0.232)		(0.012/0.245)
No answer (ref.)									
Far left				0.194*	0.194*	−0.282*	−0.282*	−0.271*	−0.271*
				(0.014/0.378)	(0.014/0.378)	(−0.541/−0.027)	(−0.541/−0.027)	(−0.484/−0.054)	(−0.484/−0.054)
Left						−0.372*	−0.372*	−0.192*	−0.192*
						(−0.603/−0.129)	(−0.603/−0.129)	(−0.380/0.000)	(−0.380/0.000)
Central									
Right - conservative	−0.118*			0.232*					0.038*
	(−0.251/−0.024)			(0.035/0.437)					(0.002/0.112)
Far right									
Non voter		−0.280*			−0.034*	−0.360*	−0.266*		0.093*
		(−0.495/−0.081)			(−0.083/−0.009)	(−0.621/−0.110)	(−0.536/0.001)		(0.024/0.209)
Trust	−0.280*	0.657*	0.945*				−0.200*		−0.326*
	(−0.561/−0.094)	(0.185/1.281)	(0.491/1.344)				(−0.431/−0.044)		(−0.580/−0.099)
R^2	0.048	0.117	0.271	0.209	0.209	0.068	0.068	0.111	0.111
Model fit	CFI=0.931; RMSEA=0.040; SRMR=0.047								

Notes: *p<0.05; unstandardized coefficients; bootstrapped confidence intervals below coefficient in parentheses; only significant coefficients are shown; negative coefficients on laws and activism indicate more acceptance of laws or more activism. N= 418.

Source: ISSP 2010, own calculations.

Table GER 7: General values and environmentalism in Germany West

	Cognitive	Evaluative	Conative	Action	Action (total)	Laws	Laws (total)	Activism	Activism (total)
Post-materialists (ref.)									
Mixed type		–0.156*	–0.099*		–0.058*		0.118*	0.164*	0.320*
		(–0.249/–0.068)	(–0.184/–0.014)		(–0.099/–0.027)		(0.066/0.186)	(0.022/0.317)	(0.177/0.484)
Materialists		–0.433*			–0.161*		0.289*	0.200*	0.539*
		(–0.591/–0.305)			(–0.239/–0.103)		(0.197/0.405)	(0.000/0.411)	(0.325/0.744)
No answer (ref.)									
Far left		–0.215*			–0.080*			–0.377*	
		(–0.437/–0.015)			(–0.179/–0.008)			(–0.768/–0.027)	
Central/left	0.101*	0.135*	0.132*		0.057*	–0.282*	–0.398*	–0.313*	–0.486*
	(0.014/0.188)	(0.059/0.227)	(0.051/0.216)		(0.028/0.104)	(–0.450/–0.129)	(–0.557/–0.251)	(–0.446/–0.179)	(–0.626/–0.353)
Central									
Right – conservative						–0.259*	–0.259*		
						(–0.419/–0.087)	(–0.419/–0.087)		
Far right				–0.529*	–0.529*				
				(–1.020/–0.042)	(–1.020/–0.042)				
Non voter		–0.121*			–0.045*		0.070*		0.069*
		(–0.226/–0.017)			(–0.093/–0.007)		(.0.004/0.147)		(0.001/0.146)
Trust	–0.382*	0.458*	0.284*		0.143*		–0.290*		–0.369*
	(–0.623/–0.161)	(0.250/0.696)	(0.070/0.483)		(0.072/0.246)		(–0.448/–0.159)		(–0.587/–0.231)
R^2	0.019	0.225	0.170	0.095	0.095	0.130	0.130	0.229	0.229
Model fit	CFI=0.957; RMSEA=0.035; SRMR=0.036								

Notes: *p<0.05; unstandardized coefficients; bootstrapped confidence intervals below coefficient in parentheses; only significant coefficients are shown; negative coefficients on laws and activism indicate more acceptance of laws or more activism. N= 989.

Source: ISSP 2010, own calculations.

The central/left voters are most concerned in all three dimensions. This category contains people who vote for the green party. However this effect is barely translated into everyday action. Interestingly, the far left are least concerned about the evaluative dimension, followed by the non-voters. The higher environmentalism of the left is expressed most strongly in this category's support for laws and activism, instead of their everyday activities. Obviously the attention is shifted to these more symbolic activities and responsibility thus given to the lawmaker and those addressed by activism. The far right opposes environmentally friendly behavior in their day-to-day routine most. Like left-wing voters, right/conservatives support laws more strongly than the rest.

To sum up, the effects point in a direction to be expected by the theoretical assumptions: environmentalism is more distinct among the left with the minor exception of more support for laws demonstrated by the conservatives and less evaluative concern by the far left.

As in virtually all other countries, in Germany West trust negatively influences the cognition of environmental problems and positively influences the two other dimensions. Because the strongest predictor of each kind of action is the evaluative environmental concern, trust likewise increases all types of action indirectly represented by the significant total effects.

In Germany West, general values can help a little to explain people's view on environmental issues (Table GER 7), especially with regard to the acceptance of laws and activism where the coefficients are somewhat larger. However, most of the effects of the value variables are already found in environmental concern. The strongest effects from the three values' representation are on the evaluative dimension. This dimension, in turn, is the strongest significant predictor of all three types of environmental action. Obviously, environmentalism is a relatively value-based issue and thus, when acting pro-environmental as a result of environmental concern, this is a value-rational decision. However, due to the low explained variance and the high amount of action people already show, environmentalism has spread into society. Virtual environmentalism, as mentioned for the former GDR, is even more present in the old FRG states.

Social positions (Table GER 8 and 9): In Germany East, like in the Czech Republic, the educational reference category represents the lowest level of formal education instead of 'no formal education' because only four people fall into the latter category. There are also only four that gave no answer to their occupational position; they are included in the reference category.

Table GER 8: Social positions and environmentalism in Germany East

	Cognitive	Evaluative	Conative	Action	Action (total)	Laws	Laws (total)	Activism	Activism (total)
Equivalence income	−0.056*								
	(−0.106/−0.019)								
Equivalence income^2									
Lowest formal education (ref.)									
Intermediate		0.233*		−0.301*	−0.309*	0.192*			
		(0.062/0.433)		(−0.478/−0.138)	(−0.482/−0.146)	(0.004/0.396)			
Secondary		0.413*		−0.608*	−0.552*	0.495*	0.353*		−0.155*
		(0.123/0.782)		(−0.908/−0.335)	(−0.852/−0.274)	(0.149/0.894)	(0.004/0.757)		(−0.325/−0.038)
University incomplete	0.194*	0.489*		−0.412*			−0.171*		−0.248*
	(0.037/0.394)	(0.161/0.844)		(−0.738/−0.111)			(−0.365/−0.055)		(−0.454/−0.101)
University complete		0.686*		−0.330*	−0.258*		−0.239*		−0.253*
		(0.454/0.961)		(−0.567/−0.106)	(−0.473/−0.047)		(−0.415/−0.099)		(−0.424/−0.111)
Never had paid work (ref.)									
No paid work									
In paid work									
No answer									
School children in household									
Toddlers in household									
Big city (ref.)									
Suburbs									
Town									
Village/Farm									
R^2	0.045	0.157	0.203	0.259	0.259	0.064	0.064	0.099	0.099
Model fit	CFI=0.938; RMSEA=0.038; SRMR=0.046								

Notes: *p<0.05; unstandardized coefficients; bootstrapped confidence intervals below coefficient in parentheses; only significant coefficients are shown; negative coefficients on laws and activism indicate more acceptance of laws or more activism. N=418.

Source: ISSP 2010, own calculations.

Table GER 9: Social positions and environmentalism in Germany West

	Cognitive	Evaluative	Conative	Action	Action (total)	Laws	Laws (total)	Activism	Activism (total)
Equivalence income		0.100*			0.027*	–0.064*	–0.114*		–0.059*
		(0.057/0.155)			(0.012/0.048)	(–0.129/.001)	(–0.181/–0.051)		(–0.090/–0.028)
Equivalence income²		–0.012*							
		(–0.021/–0.004)							
No formal education (ref.)									
Lowest formal education				0.244*	0.244*				
				(0.109/0.383)	(0.109/0.383)				
Intermediate		0.146*		0.180*	0.249*		–0.129*		–0.102*
		(0.063/0.245)		(0.054/0.328)	(0.112/0.403)		(–0.203/–0.069)		(–0.169/–0.048)
Secondary		0.311*	0.119*		0.292*		–0.258*	–0.269*	–0.537*
		(0.167/0.457)	(–0.001/0.244)		(0.086/0.480)		(–0.380/–0.165)	(–0.490/–0.051)	(–0.765/–0.306)
University incomplete		0.429*		0.244*	0,428*		–0.395*	–0.650*	–0.966*
		(0.255/0.634)		(0.011/0.440)	(0.196/0.666)		(–0.565/–0.271)	(–0.926/–0.347)	(–1.251/–0.665)
University complete		0.488*	0.275*		0.216*		–0.469*	–0.237*	–0.724*
		(0.311/0593)	(0.166/0.400)		(0.138/0.319)		(–0.604/–0.354)	(–0.450/–0.048)	(–0.925/–0.519)
Never had paid work (ref.)									
No paid work	–0.081*	0.137*			0.050*		–0.080*		
	(–0.174/–0.001)	(0.012/0.266)			(0.002/0.116)		(–0.172/–0.004)		
In paid work		0.176*			0.082*		–0.164*	–0.163*	–0.293*
		(0.056/0.297)			(0.034/0.146)		(–0.261/–0.086)	(–0.286/–0.031)	(–0.436/–0.155)
No answer						0.643*	0.643*		
						(0.110/1.202)	(0.110/1.202)		
School children in household					–0.004*		0.018*		0.009*
					(–0.009/–0.002)		(0.008/0.036)		(0.004/0.018)
Toddlers in household					–0.007*		0.030*		0.015*

Table GER 9: Continued

	Cognitive	Evaluative	Conative	Action	Action (total)	Laws	Laws (total)	Activism	Activism (total)
					(−0.015/−0.003)		(0.012/0.056)		(0.007/0.029)
Big city (ref.)									
Suburbs									
Town			0.122*				−0.020*		−0.069*
			(0.045/0.203)				(−0.051/−0.003)		(−0.122/−0.027)
Village		0.124*		0.104*	0.160*		−0.094*		−0.083*
		(0.047/0.206)		(0.006/0.191)	(0.068/0.250)		(−0.157/−0.043)		(−0.139/−0.033)
Farm									
R^2	0.004	0.271	0.195	0.119	0.119	0.132	0.132	0.230	0.230
Model fit	CFI=0.980; RMSEA=0.025; SRMR=0.033								

Notes: *p<0.05; unstandardized coefficients; bootstrapped confidence intervals below coefficient in parentheses; only significant coefficients are shown; negative coefficients on laws and activism indicate more acceptance of laws or more activism. N= 989.

Source: ISSP 2010, own calculations.

Finally, only three persons state to be living on a farm in the country. They are combined with the 'village' category.

For Germany East the impact of income is quite straightforward (Table GER 8): richer people are less environmentally concerned in the cognitive dimension than those with a lower income. This might reflect the fact that more wealthy people are less exposed to environmental pollution. The other dimensions as well as any type of action are not influenced by the equivalence income. None of the coefficients of the squared equivalence income is significant, neither on concern nor on any type of action.

The strongest effects are found among the education variables, but the signs of the coefficients are contradictory, at least at first sight. The evaluative dimension in particular is strongly and positively influenced by education. Expressly those with a university degree stand out. With regard to the cognitive dimension only those with an incomplete university degree (who are mainly students) are more concerned than people holding another degree, either higher or lower.

Striking is that holding an educational degree other than the countries lowest degree entails significantly lower participation in pro-environmental everyday action. Among people with an intermediate or higher secondary degree, the acceptance of laws is also less pronounced. Why is it that those who are much more concerned are least active? First, this is a strong articulation of the discrepancy between aspiration and reality. People hold strong evaluative attitudes, which may be mainly expressed verbally or symbolically: Only the acceptance of laws and activism are affected by the evaluative concern which transfers the effect of education. People shift their attention to more visible activities. When it comes to everyday action, other problems have priority.

Further support for the priority assumption comes from Table GER 1 where the economy, health care, education and poverty rank above the environment. Thus, people may be more engaged in activities relating to these issues rather than the environment but this suggestion remains unresolved and open to further research.

Results of the occupational position, the number of toddlers and children in the household as well as the size of the community the respondents live in, are unspectacular: These aspects neither contribute to environmental concern nor to any type of action; none of the coefficients are significant.

Turning to Germany West, only five respondents gave no answer concerning their education and are included in the reference category. All respondents answered the question about the community size substantively and no dummy for the 'no answer' category is needed.

Table GER 9 summarizes the results for Germany West. As is the case in Germany East the coefficients of the equivalence income are straightforward, yet point in the opposite direction. While in the East income has its only significant effect (which is negative) on the cognitive dimension, in the West the evaluative dimension is influenced positively. Concerning non-linearity, the introduction of the squared equivalence income only yielded a significant coefficient on the evaluative dimension indicating a concave trajectory, i.e. first, a higher income increases the evaluation of environmental concern, then after a turning point it decreases with increasing income.

Furthermore, the effects of education are more in line with the theoretical expectations than those of Germany East. People holding an intermediate, higher secondary, incomplete or complete university degree are more environmentally concerned on the evaluative dimension. Moreover university graduates are more willing to do something than those with just a higher secondary education. All five educational categories exhibit higher (total) values in everyday action in contrast to those without formal education. This kind of behavior is most pronounced among the incomplete university group. The more or less equal coefficients of the other groups result from the combination of the direct effects of lowest and intermediate education on action and the stronger indirect effects of a secondary and university degree. The acceptance of laws is not affected by education directly but mediated via environmental concern; a higher education entails an increasing acceptance of laws. Environmental activism is additionally influenced directly by having an incomplete university degree (most likely students) leading to the highest values.

In contrast to Germany East, at this point of analysis, the occupational position plays a certain role in Western Germany. Having no paid job at the moment as well as having an occupation leads to higher evaluative concern. Thus the significant total effects on action can be explained. Given that for activism the cognitive dimension also shows a significant coefficient (see Table 5.4) and this dimension, in turn, is negatively impacted by not having a paid job, there is no total effect of having no job on activism. In particular the total effects of having a job (plus the direct effect on activism) runs counter to the biographic availability assumption that having an occupation leaves less time for being environmentally friendly. The findings instead support the counter argument that a more stable situation allows people to deal with other less acute and personal issues.

Table GER 10: Full model Germany East

	Cognitive	Evaluative	Conative	Action	Action (total)	Laws	Laws (total)	Activism	Activism (total)
Gender		–0.134*			–0.016*	0.169*	0.208*		0.043*
		(–0.269/0.000)			(–0.053/–0.001)	(0.000/0.359)	(0.031/0.391)		(0.000/0.111)
Born before 1934 (ref.)									
1934–1943									
1944–1953									
1954–1963									
1964–1973		0.315*			0.039*		–0.088*		–0.104*
		(0.095/0.506)			(0.010/0.091)		(–0.192/–0.019)		(–0.215/–0.037)
1974–1983				–0.431*	–0.431*			0.279*	0.279*
				(–0.629/–0.222)	(–0.629/–0.222)			(0.124/0.423)	(0.124/0.423)
Born after 1983				–0.357*	–0.346*	0.671*	0.671*		–0.004*
				(–0.574/–0.170)	(–0.558/–0.163)	(0.329/10.009)	(0.329/10.009)		(–0.017/–0.001)
Post-materialists (ref.)									0.055*
Middle		–0.150*			–0.018*		0.043*		(0.004/0.138)
		(–0.316/0.000)			(–0.054/–0.002)		(0.004/0.121)		0.045*
Materialists			–0.215*		–0.062*				(0.008/0.112)
			(–0.421/–0.025)		(–0.154/–0.009)				
No answer (ref.)									
Far left								–0.250*	–0.250*
								(–0.496/–0.013)	(–0.496/–0.013)
Left						–0.225*	–0.225*	–0.186*	–0.186*
						(–0.389/–0.054)	(–0.389/–0.054)	(–0.387/0.000)	(–0.387/0.000)
Central									
Right/Conservative	–0.101*				–0.084*				
	(–0.220/–0.005)				(–0.169/–0.021)				

Table GER 10: Continued

	Cognitive	Evaluative	Conative	Action	Action (total)	Laws	Laws (total)	Activism	Activism (total)
Far right									
Non-voter			−0.217*		−0.063*				0.047*
			(−0.385/−0.044)		(−0.140/−0.016)				(0.013/0.102)
Trust	−0.227*	0.519*	0.931*				−0.132*		−0.286*
	(−0.519/−0.008)	(0.068/10.043)	(0.490/10.406)				(−0.312/−0.023)		(−0.529/−0.087)
Equivalence income	−0.046*				−0.034*				
	(−0.091/−0.015)				(−0.068/−0.012)				
Equivalence income²									
Lowest formal education (ref.)									
Intermediate				−0.262*	−0.246*		−0.049*		−0.062*
				(−0.431/−0.083)	(−0.409/−0.063)		(−0.151/−0.001)		(−0.153/−0.001)
Higher secondary		0.344*		−0.405*	−0.362*		−0.093*		−0.120*
		(0.028/0.654)		(−0.683/−0.159)	(−0.639/−0.100)		(−0.273/−0.015)		(−0.254/−0.021)
University incomplete	0.183*	0.439*		−0.454*	−0.293*		−0.127*		−0.201*
	(0.018/0.412)	(0.131/0.720)		(−0.772/−0.172)	(−0.588/0.000)		(−0.292/−0.025)		(−0.365/−0.084)
University		0.589*		−0.309*	−0.271*		−0.171*		−0.188*
		(0.381/0.832)		(−0.539/−0.118)	(−0.488/−0.068)		(−0.347/−0.053)		(−0.343/−0.072)
Never had paid work (ref.)									
No paid work									
In paid work					−0.018*				0.007*
					(−0.040/−0.006)				(0.001/0.023)
NA work									
School children in household					0.005*				−0.002*

Table GER 10: Continued

	Cognitive	Evaluative	Conative	Action	Action (total)	Laws	Laws (total)	Activism	Activism (total)
					(0.001/0.015)				(–0.008/0.000)
Toddlers in household					0.008*				–0.003*
					(0.002/0.020)				(–0.010/0.000)
Big city (ref.)									
Suburbs					–0.012*				0.005*
					(–0.034/–0.002)				(0.000/0.019)
Town					–0.011*				0.004*
					(–0.029/–0.002)				(0.001/0.014)
Village					–0.014*				0.006*
					(–0.039/–0.004)				(0.001/0.018)
Farm									
Cognitive				0.738*	0.738*			–0.300*	–0.300*
				(0.408/10.272)	(0.408/10.272)			(–0.660/–0.035)	(–0.660/–0.035)
Evaluative			0.424*		0.123*	–0.298*	–0.298*	–0.248*	–0.339*
			(0.168/0.676)		(0.051/0.236)	(0.–0.685/–0.104)	(0.–0.685/–0.104)	(–0.482/–0.041)	(–0.573/–0.146)
Conative				0.289*	0.289*			–0.214*	–0.214*
				(0.169/0.447)	(0.169/0.447)			(–0.319/–0.099)	(–0.319/–0.099)
R^2	0.078	0.240	0.261	0.324	0.324	0.109	0.109	0.128	0.128
Model fit	CFI=0.963; RMSEA=0.030; SRMR=0.000								

Notes: *p<0.05; unstandardized coefficients; bootstrapped confidence intervals below coefficient in parentheses; only significant coefficients are shown; negative coefficients on laws and activism indicate more acceptance of laws or more activism. N= 418.

Source: ISSP 2010, own calculations.

Table GER 11: Full model Germany West

	Cognitive	Evaluative	Conative	Action	Action (total)	Laws	Laws (total)	Activism	Activism (total)
Gender	0.096*		−0.101*		0.037*	0.128*			
	(0.010/0.176)		(−0.184/−0.037)		(0.009/0.073)	(0.000/0.249)			
Born before 1934 (ref.)									
1934–1943		0.185*		0.189*	0.264*		−0.100*		−0.099*
		(0.065/0.335)		(0.054/0.329)	(0.123/0.412)		(−0.202/−0.032)		(−0.199/−0.038)
1944–1953	−0.127*	0.289*			0.107*		−0.115*		−0.125*
	(−0.235/−0.017)	(0.165/0.427)			(0.051/0.172)		(−0.216/−0.035)		(−0.228/−0.053)
1954–1963		0.388*			0.157*		−0.212*		−0.216*
		(0.268/0.547)			(0.099/0.243)		(−0.340/−0.136)		(−0.359/−0.123)
1964–1973		0.266*			0.148*		−0.203*		−0.201*
		(0.247/0.504)			(0.092/0.237)		(−0.318/−0.124)		(−0.331/−0.121)
1974–1983	0.158*	0.271*			0.118*		−0.174*		−0.172*
	(0.023/0.287)	(0.136/0.430)			(0.062/0.202)		(−0.294/−0.086)		(−0.298/−0.093)
Born after 1983		0.175*		−0.288*	−0.233*	0.372*	0.302*		−0.087*
		(0.034/0.338)		(−0.415/−0.138)	(−0.357/−0.058)	(0.157/0.607)	(0.070/0.522)		(−0.199/−0.013)
Post-materialists (ref.)									
Middle		−0.098*			−0.040*		0.058*		0.062*
		(−0.176/−0.001)			(−0.078/−0.005)		(0.011/0.114)		(0.014/0.126)
Materialists		−0.355*			−0.147*		0.215*		0.210*
		(−0.489/−0.245)			(−0.220/−0.088)		(0.142/0.324)		(0.125/0.344)
No answer (ref.)									
Far left		−0.246*			−0.106*		0.150*	−0.398*	
		(−0.459/−0.056)			(−0.217/−0.028)		(0.049/0.266)	(−0.839/−0.005)	

Table GER 11: Continued

	Cognitive	Evaluative	Conative	Action	Action (total)	Laws	Laws (total)	Activism	Activism (total)
Left	0.103*	0.109*	0.127*		0.052*	−0.299*	−0.384*	−0.308*	−0.450*
	(0.019/0.204)	(0.024/0.194)	(0.051/0.207)		(0.018/0.098)	(−0.453/−0.156)	(−0.537/−0.244)	(−0.454/−0.168)	(−0.605/−0.298)
Central			−0.154*						0.079*
			(−0.241/−0.074)						(0.035/0.139)
Right/Conservative					−0.002*	−0.261*	−0.250*		0.003*
					(−0.008/0.000)	(−0.431/−0.100)	(−0.415/−0.081)		(0.000/0.011)
Far right				−0.470*	−0.470*				
				(−0.807/−0.069)	(−0.807/−0.069)				
Non-voter		−0.114*			−0.051*		0.085*		0.067*
		(−0.223/−0.012)			(−0.095/−0.009)		(0.024/0.150)		(0.016/0.134)
Trust	−0.361*	0.265*	0.206*		0.090*				−0.221*
	(−0.595/−0.130)	(0.061/0.466)	(0.011/0.435)		(0.009/0.177)				(−0.388/−0.076)
Equivalence income		0.049*			0.020*	−0.063*	−0.091*		−0.029*
		(0.011/0.098)			(0.005/0.044)	(−0.133/0.000)	(−0.159/−0.031)		(−0.062/−0.009)
Equivalence income²									
No formal education (ref.)									
Lowest formal education		−0.117*		0.208*	0.155*		0.090*		0.068*
		(−0.208/−0.003)		(0.079/0.347)	(0.031/0.280)		(0.045/0.153)		(0.025/0.135)
Intermediate				0.194*	0.194*				
				(0.075/0.333)	(0.075/0.333)				
Higher secondary			0.122*	0.241*	0.241*			−0.248*	−0.311*
			(0.018/0.279)	(0.089/0.399)	(0.089/0.399)			(−0.549/0.001)	(−0.635/−0.050)

Table GER 11: *Continued*

	Cognitive	Evaluative	Conative	Action	Action (total)	Laws	Laws (total)	Activism	Activism (total)
University incomplete		0.190*		0.213*	0.300*		–0.150*	–0.638*	–0.758*
		(0.039/0.355)		(0.127/0.430)	(0.106/0.505)		(–0.251/–0.061)	(–10.049/–0.313)	(–10.176/–0.422)
University		0.236*	0.271*		0.110*		–0.192*	–0.237*	–0.524*
		(0.116/0.367)	(0.127/0.430)		(0.054/0.177)		(–0.283/–0.115)	(–0.487/–0.007)	(–0.780/–0.288)
Never had paid work (ref.)									
No paid work									
In paid work					0.007*		–0.032*	–0.179*	–0.189*
					(0.002/0.015)		(–0.058/–0.011)	(–0.298/–0.059)	(–0.307/–0.067)
NA work									
School children in household					–0.003*		0.016*		0.005*
					(–0.008/–0.001)		(0.007/0.028)		(0.002/0.011)
Toddlers in household					–0.004*		0.019*		0.006*
					(–0.010/–0.001)		(0.005/0.038)		(0.002/0.014)
Big city (ref.)									
Suburbs									
Town			0.136*						–0.072*
			(0.056/0.217)						(–0.133/–0.023)
Village		0.117*		0.111*	0.161*		–0.072*		–0.068*
		(0.046/0.200)		(0.101/0.203)	(0.063/0.253)		(–0.119/–0.028)		(–0.124/–0.026)
Farm			0.174*						–0.092*
			(0.000/0.513)						(–0.291/0.002)

Table GER 11: Continued

	Cognitive	Evaluative	Conative	Action	Action (total)	Laws	Laws (total)	Activism	Activism (total)
Cognitive		0.202*	0.139*		0.082*	–0.159*	–0.267*		–0.180*
		(0.122/0.296)	(0.061/0.237)		(0.044/0.131)	(–0.287/–0.031)	(–0.391/–0.144)		(–0.275/–0.113)
Evaluative			0.273*	0.405*	0.405*	–0.537*	–0.537*	–0.409*	–0.541*
			(0.157/0.425)	(0.270/0.575)	(0.270/0.575)	(–0.743/–0.373)	(–0.743/–0.373)	(–0.646/–0.214)	(–0.776/–0.342)
Conative								–0.504*	–0.504*
								(–0.699/–0.300)	(–0.699/–0.300)
R^2	0.040	0.385	0.227	0.161	0.161	0.155	0.155	0.255*	0.255
Model Fit	CFI=0.983; RMSEA=0.021; SRMR=0.024								

Notes: *p<0.05; unstandardized coefficients; bootstrapped confidence intervals below coefficient in parentheses; only significant coefficients are shown; negative coefficients on laws and activism indicate more acceptance of laws or more activism. N=989.

Source: ISSP 2010, own calculations.

For those without a job this is not that clear. A look into the data shows that those without a job are mostly low-level or intermediately educated and have (naturally) lower incomes. They also have an average age of 61 years indicating that many people falling into that category are already retired. As this category implies they had a job before, this generally stable life situation might contribute to their higher environmental concern.

In accordance with the biographic availability assumption are the total effects of having children and toddlers in the household. These effects are very small, though, and they are mediated via income. As income positively impacts the evaluative environmental concern and at the same time children/toddlers lead to a lower household income, children or toddlers decrease all three types of action.

People living in a village are more environmentally concerned (evaluative dimension), consequently also more active in all three aspects. If living in a town, people are a bit more willing to pay. This effect is only mediated to the acceptance of laws and activism because everyday action is not affected by the conative dimension of concern.

Full model (Tables GER 10 and GER 11): In Germany East (Table GER 10) major differences in the full model are declining coefficients linked with the inclusion of additional parameters. Most notable seem to be the disappearing coefficients of the far-left political orientation on action and the acceptance of laws when the social positions are included. At the same time, the total effects of having an incomplete or complete university degree increase in size and for the former (incomplete) this becomes significant in the full model while the latter (complete) remains significant. This indicates that under control of values, cohorts and gender, the negative direct effects on action are less strongly balanced by the positive effects on concern than in the model restricted to social positions.

Despite the very small coefficient sizes in the full model the total effects of having a paid occupation, children or toddlers in household as well as community size become significant. In accordance with the biographic availability assumption, a job or children/toddlers minimally decrease the engagement in everyday action and activism. Likewise, people living in suburbs, a town or village are less active than people from other community types in their day-to-day routine and activism mediated via income.[17]

17 Recall from 4.2.2 that due to its metric properties income is also the dependent variable for the other independent variables thus it can mediate the effects of the community size.

Basically, in Germany West (Table GER 11) the same happens when including all factors: most coefficients decrease in size, some thereby lose significance, and a few increase in size (e.g. gender on laws or $cohort_{1984+}$ on laws). Some notable differences are mentioned nevertheless. While the intermediate and higher secondary educated no longer show significant effects, those with lowest education are now slightly less concerned on the evaluative dimension than in the model where only social positions are included (Table GER 9). At the same time the direct effects of materialism and mixed type on activism lose significance which as a result entails smaller total effects. While in Table GER 9 there were significant effects of the occupational situation of the evaluative and cognitive dimensions, these effects disappear once controlling for cohort effects.

The analysis of Germany shows that from an ecological perspective there are still substantial differences between the two parts of the country. The descriptive analysis and the salience of environmental issues instead indicate a small ecology deficit in the amount of environmentalism; the multivariate analyses underline cultural or historical differences in the formation of environmentalism. These differences may well be a result of the ecology deficit in Germany East in the past two decades following the German reunification. Individual environmentalism needed to develop as some kind of innovation driven for instance by values or social position. The convergence of East and West it would suggest an assumption that both the degree of environmentalism will settle down to a more or less equally high level and the amount of variance attributed to socially- or value-based traits might decline in Germany East. The cue 'cultural difference' directly leads on now to the next country. Covering an approximately 18,000 km linear distance from Germany to New Zealand, a liberal welfare state in another part of the world is spotlighted next.

6.3 New Zealand

GDP per capita (2010): 32,407 $

In their study with the ISSP data of 1993, Dekker et al. (1997) ranked New Zealand 7th out of 20 countries regarding its citizens' environmental concern (Dekker et al. used the same measures that represent the cognitive dimension in the present study). However, it shared this rank with Northern Ireland. Additionally only 11 ranking positions could be given because many countries hold the same mean (Dekker *et al.* 1997, p. 449). The New Zealanders' willingness to pay ranked up to 5th out of 20, sharing the position with the Japanese and US-American respondents. Due to several equal means, only 14 ranking positions can be distributed. A similar position with the same data but different measures is attributed to New

Zealand by Diekmann and Franzen (1999) where it occupied sixth place together with Germany West and ranked sixth for willingness to pay as found in Kemmelmeyer et al. (2002). In their six-nations comparison, Schultz et al. (2005) found the second highest amount of environmental behavior among the New Zealander's non-representative student sample, behind Germany which took first position. Concerning the seriousness of local or global environmental problems, New Zealand and Germany ranked 5th and 4th, respectively. In ISSP 2000 data, New Zealand is found in 8th position alongside Austria out of 26 countries (Franzen and Meyer 2010). This very brief overview places New Zealand's environmental concern in an upper middle position. Unfortunately, in the Inglehart (1995) study New Zealand was not part of the World Values Survey. Concerning environmental behavior, Hadler and Haller (2011) found New Zealand ranked eleventh (out of 23) in private behavior (recycling and cutting back car driving) but ranked two in public behavior (the same indicators as used here). In the present study New Zealand (Table 5.1) ranks third lowest in the cognitive and evaluative dimension which tends to be lower than in the cases cited above. Regarding New Zealand's willingness to pay, the present results resemble past findings as New Zealand is ranked seventh. Likewise, for everyday action or the acceptance of laws the country can be found in position 14 and for activism it ranks second behind Switzerland. Assessing whether New Zealanders think or act environmentally consciously or not depends very much on the aspect considered.

In New Zealand the relationship between the three dimensions of environmental concern is relatively strong as is their impact on everyday action. This impact is strongest among the 19 countries looked at in this study. The cognition's effect is most intense, followed that of the evaluative and finally that of conative. They together explain 38% of the variance in everyday action in New Zealand. Environmental concern accounts for 19.9% and 20.7% respectively in terms of acceptance of laws and activism. In New Zealand one can clearly not yet speak of virtual environmentalism but rather of both value- and instrumental-rational environmental action.

The order of importance of various issues in Table NZ 1 is almost exactly the same order as the average across all six nations (see Table 6.1). Only the last two categories changed their position. In this regard New Zealand seems to be the most prototypical country while the other nations deviate to a greater or lesser extent from this order of importance. This also entails the typical middle position of the environment in relation to other aspects.

Table NZ 1: Most important issue in New Zealand

	Most important	Second most important	Score
The economy	365	158	444
Health care	283	282	424
Education	166	297	314.5
Crime	160	150	235
The environment	97	147	170.5
Poverty	41	52	67
Immigration	13	39	32.5
None of these	7	5	9.5
Terrorism	1	5	3.5

Notes: Source: ISSP 2010, own calculations.

A look at Table NZ 2 reveals an interesting irregularity. Across all other six nations air pollution is among the three most important issues and in New Zealand this does not seem to be an overly serious issue. Instead, the top three is occupied by water pollution, chemicals and pesticides (which may pollute the water) and domestic waste disposal. This finding underlines the local influences not only on how environmentally concerned people are but also that issues seen as a problem can be quite versatile in different geographic regions. The fourth-place ranking of climate change seems pretty surprising as the New Zealand Economic and Financial Overview (New Zealand Government 2013, p. 7) states "climate change presents a particular challenge for New Zealand, both from an international and domestic perspective". For the respondents local and contemporary problems are more prevalent.

In New Zealand the share of solar, wind and water energy is highest among the six countries investigated in this study (Table NZ 3). 32.5% of its primary energy already stems from these sources, which are also considered as most important by the respondents. Fossil fuels still play the most important role in the actual situation but only rank third in the opinion of people. Despite the fact that nuclear energy does not contribute to New Zealand's energy supply at all, 10.9% consider it the best source for the future and thereby it is ranked second.

Table NZ 2: Most important environmental issue in New Zealand

	For country	For family	Score
Water pollution	300	138	219
Chemicals and pesticides	178	173	176
Domestic waste disposal	138	186	162
Climate change	133	96	115
Using up our natural resources	135	77	106
None of these	18	158	88
Water shortage	87	66	77
Air pollution	51	85	68
Genetically modified foods	51	83	67
Nuclear waste	7	6	7

Notes: Source: ISSP 2010, own calculations.

Table NZ 3: Energy supply in New Zealand

	Best source for future energy supply		Real total primary energy supply (TPES) 2010
	%	N	%
Solar, wind or water power	78.7%	852	32.5%
Nuclear power	10.3%	111	0.0%
Coal, oil and natural gas	7.9%	85	60.8%
Fuels made from crops	2.5%	27	6.4%
None of these	0.6%	7	0.2%
Total	100%	1082	100%

Notes: Source: ISSP 2010, own calculations. International Energy Agency (2012).

Multivariate analysis

As with Germany East the fit of the basic model is not optimal (CFI=0.928; RMSEA=0.068; SRMR=0.063). Rather than low factor loadings, the issue in terms of New Zealand concerns various implied relations between the observed variables that go beyond what is accounted for by the relation between the latent constructs. For example, a small correlation between the residuals of Q20d (save energy) and Q20e (save water) is implied by modification indices. This suggests that these two variables have something in common not accounted for in the latent variable of everyday action. They are both saving behavior while the third variable Q20f is consumption behavior. However, the implied parameter is small than 0.1. Several such minor residual correlations can be found. The accumulation of many minor misspecifications contributes to the suboptimal model fit. Again, for comparability the model is retained because the fit indices indicate an at least acceptable model fit. Cognitive → affective=0.454 (0.374/0.535); cognitive → conative=0.176 (0.117/0.235); affective → conative=0.234 (0.179/0.304).

Table NZ 4: Cohorts, gender and environmentalism in New Zealand

	Cognitive	Evaluative	Conative	Action	Action (total)	Laws	Laws (total)	Activism	Activism (total)
Male (ref.)									
Female	0.255*			0.129*	0.236*	0.145*			−0.112*
	(0.123/0.314)			(0.039/0.225)	(0.139/0.333)	(0.025/0.259)			(−0.171/−0.065)
Born before 1934 (ref.)									
1934–1943									
1944–1953	0.230*						−0.109*	−0.154*	−0.226*
	(0.076/0.381)						(−0.193/−0.037)	(−0.313/0.002)	(−0.417/−0.041)
1954–1963	0.208*		−0.189*				−0.158*		
	(0.057/0.378)		(−0.340/−0.032)				(−0.250/−0.061)		
1964–1973	0.313*	0.206*	−0.231*		0.102*		−0.248*		−0.197*
	(0.159/0.469)	(0.072/0.334)	(−0.395/−0.073)		(0.005/0.203)		(−0.349/−0.144)		(−0.324/−0.073)
1974–1983	0.333*	0.336*	−0.250*		0.126*	0.259*			−0.268*
	(0.164/0.503)	(0.186/0.483)	(−0.414/−0.089)		(0.022/0.234)	(0.093/0.429)			(−0.407/−0.132)
Born after 1983	0.209*	0.180*	−0.177*	−0.169*		0.370*			−0.151*
	(0.014/0.387)	(0.005/0.352)	(−0.354/−0.003)	(−0.312/0.011)		(0.189/0.579)			(−0.294/−0.020)
Cognitive									
Evaluative									
Conative									
R^2	0.045	0.230	0.248	0.350	0.350	0.219	0.219	0.209	0.209
Model fit	CFI=0.942; RMSEA=0.50; SRMR=0.046								

Notes: *p<0.05; unstandardized coefficients; bootstrapped confidence intervals below coefficient in parentheses; only significant coefficients are shown; negative coefficients on laws and activism indicate more acceptance of laws or more activism. N=1172.

Source: ISSP 2010, own calculations.

Cohorts and gender (NZ 4): The gender-environment relation in New Zealand can be described as follows: women are more environmental friendly in New Zealand (cognitive and action). They also are less likely to accept laws, and in combination with the indirect effect on laws the total effect is not significant. The total effects on action and activism, however, are significant.

The cohort effects follow an inverted U-shaped curve. Lowest concern (cognitive and evaluative) is found in the oldest cohorts. The peak is present in the 1964–73 and 1974–83 cohorts and then the coefficient size declines again with the youngest generation. Interestingly, the same tendency is true for the willingness to pay only that the signs of the coefficients are negative. Again, those who are most concerned are least willing to do something. This poses a conflict as the cognitive and conative dimension both contribute to everyday action in New Zealand thus the two coefficients, to some extent, counteract each other and thus reduce the mediated effect on action. As can be seen, only the total effects of the two most concerned generations (1964–73 and 1974–83) remain significant. Only the 1984+ cohort is directly less active. Laws are least accepted by the two youngest cohorts in the sample and in total the cohorts from 1944–1963 support laws more strongly. The 1944–1953 cohort is most often engaged in activism. When taking indirect effects via environmental attitudes into account, only the coefficients of the 1954–63 are non-significant because the positive effect of the cognitive and the negative effect of the conative dimension counteract each other.

General values (NZ 5): The way in which general values affect environmental concern is relatively straightforward. Post-materialists perceive environmental problems as slightly more severe and are more willing to pay than those in the middle position or materialists. There are no differences concerning the evaluative dimension. The total negative effects of both categories on action are significant. There is an interesting negative coefficient on the acceptance of laws, which indicated that materialists accept laws more willingly than post-materialists or mixed types. An explanation for this instance is already mentioned for the Czech Republic (general support for laws, attention shifting; see profile of the Czech Republic). As this a recurring finding, this issues will be discussed in the summary of the finding in chapter 7.2.2. In total, these effects are countered by the lower concern in this group. With regard to activism the expected indications are found: mixed types and materialists are less active to a relatively large extent. The role of the individual's political position is more intuitive as for instance in the Czech Republic, yet this does not justify talk of a linear relation.

Table NZ 5: General values and environmentalism in New Zealand

	Cognitive	Evaluative	Conative	Action	Action (total)	Laws	Laws (total)	Activism	Activism (total)
Post-materialists (ref.)									
Mixed type	−0.197*		−0.252*		−0.178*			0.416*	0.565*
	(−0.384/−0.011)		(−0.376/−0.130)		(−0.280/−0.083)			(0.173/0.657)	(0.304/0.807)
Materialists	−0.179*		−0.337*		−0.199*	−0.258*		0.551*	0.711*
	(−0.384/0.002)		(−0.470/−0.202)		(−0.312/−0.092)	(−0.511/−0.006)		(0.280/0.826)	(0.423/0.981)
No answer (ref.)									
Far left	0.311*	0.806*	0.182*	0.273*	0.539*		−0.552*	−0.621*	−1.189*
	(0.110/0.514)	(0.587/1.017)	(0.047/0.331)	(0.073/0.462)	(0.330/0.747)		(−0.727/−0.401)	(−0.906/−0.338)	(−1.484/−0.898)
Central/left		0.190*			0.014*		−0.220*		−0.092*
		(0.052/0.322)			(0.004/0.028)		(−0.389/−0.066)		(−0.172/−0.033)
Central		0.683*			0.049*		−0.310*		−0.313*
		(0.216/1.146)			(0.014/0.099)		(−0.577/−0.096)		(−0.578/−0.085)
Right/conservative	−0.286*	0.125*			−0.125*	−0.261*	−0.172*		
	(−0.395/−0.178)	(0.007/0.244)			(−0.185/−0.070)	(−0.387/−0.136)	(−0.307/−0.031)		
Far right	−0.589*				−0.275*		0.291*		0.262*
	(−0.945/−0.288)				(−0.462/−0.132)		(0.127/0.480)		(0.111/0.442)
Non voter	−0.208*	0.204*							
	(−0.409/−0.018)	(0.009/0.402)							
Trust	−0.331*	0.497*	0.191*			−0.389*	−0.467*	0.374*	
	(−0.585/−0.113)	(0.278/0.722)	(0.019/0.348)			(−0.667/−0.080)	(−0.766/−0.133)	(0.072/0.709)	
R^2	0.072	0.288	0.268	0.345	0.345	0.222	0.222	0.237	0.237
Model fit	CFI=0.943; RMSEA=0.046; SRMR=0.043								

Notes: *p<0.05; unstandardized coefficients; bootstrapped confidence intervals below coefficient in parentheses; only significant coefficients are shown; negative coefficients on laws and activism indicate more acceptance of laws or more activism. N=1172.

Source: ISSP 2010, own calculations.

Table NZ 6: Social positions and environmentalism in New Zealand

	Cognitive	Evaluative	Conative	Action	Action (total)	Laws	Laws (total)	Activism	Activism (total)
Equivalence income	−0.095*	0.095*			−0.032*	−0.073*	−0.067*		
	(−0.147/−0.045)	(0.046/0.143)			(−0.062/−0.006)	(−0.132/−0.011)	(−0.133/−0.003)		
Equivalence income^2								−0.063*	−0.063*
								(−0.110/−0.018)	(−0.110/−0.018)
Lowest education (ref.)									
Intermediate		0.314*			0.039*		−0.147*		−0.392*
		(0.134/0.508)			(0.005/0.092)		(−0.245/−0.068)		(−0.671/−0.128)
Secondary	0.117*	0.370*			0.101*		−0.392*	−0.215*	−0.451*
	(−0.002/0.250)	(0.217/0.540)			(0.037/0.183)		(−0.578/−0.208)	(−0.421/0.002)	(−0.665/−0.321)
University incomplete		0.343*			0.037*	−0.169*	−0.339*	−0.212*	−0.374*
		(0.217/0.540)			(0.004/0.089)	(−0.324/−0.013)	(−0.487/−0.175)	(−0.397/−0.028)	(−0.560/−0.185)
University complete	0.215*	0.687*	0.078*		0.210*	−0.224*	−0.657*	−0.203*	−0.664*
	(0.094/0.346)	(0.531/0.852/)	(−0.002/0.165)		(0.124/0.311)	(−0.397/−0.034)	(−0.845/−0.477)	(−0.415/.003)	(−0.890/−0.455)
No answer									
No paid work (ref.)									
In paid work	0.169*		−0.122*				−0.124*		
	(0.047/0.282)		(−0.186/−0.056)				(−0.191/−0.060)		
No answer	0.232*				0.144*		−0.113*		−0.125*
	(0.052/0.421)				(0.027/0.208)		(−0.204/−0.026)		(−0.227/−0.030)
School children in household			−0.030*				0.016*		
			(−0.063/0.001)				(0.001/0.033)		
Toddlers in household									

Table NZ 6: Continued

	Cognitive	Evaluative	Conative	Action	Action (total)	Laws	Laws (total)	Activism	Activism (total)
Big city (ref.)									
Suburbs	0.108*				0.053*		−0.052*		−0.055*
	(−0.002/0.211)				(0.000/0.106)		(−0.107/0.001)		(−0.114/0.001)
Town									
Village									
Farm	−0.158*			0.193*			0.075*	−0.327*	−0.246*
	(−0.314/0.001)			(0.055/0.341)			(0.002/0.155)	(−0.547/−0.126)	(−0.475/−0.037)
R^2	0.042	0.309	0.257	0.346	0.346	0.212	0.212	0.217	0.217
Model fit	CFI=0.942; RMSEA=0.041; SRMR=0.039								

Notes: *p<0.05; unstandardized coefficients; bootstrapped confidence intervals below coefficient in parentheses; only significant coefficients are shown; negative coefficients on laws and activism indicate more acceptance of laws or more activism. N=1172.

Source: ISSP 2010, own calculations.

In all three dimensions of concern as well as action and activism, people holding far-left positions are more environmentally friendly, to the greatest degree in the evaluative dimension where also the moderately left and center oriented are more concerned. The conservatives are somewhat more concerned in this respect, too.

This is especially striking because they register the second-lowest perception of environmental threats as dangerous behind far-right voters. Obviously there is more to being conservative and this group's relationship to the environment in New Zealand than one would expect from the theoretical assumptions. Additionally, conservatives support laws strongest, at least regarding the direct effect.

The total effects resemble the direct effects and effect mediated via concern. The far left are most active in all three types, followed by the central, then the left. The far-right act least environmentally friendly.

In comparison to the other country samples, the influence of trust is less pronounced; the tendency is the same, though. Trustful people consider environmental problems to lesser degree severe but are more evaluative concerned and only differ slightly from less trustful in terms of their willingness to pay. Significant total effects only appear with regard to the acceptance of laws due to the direct effect. For action and activism, the positive and negative coefficients on environmental attitudes cancel each other out. The effects of environmental attitudes even override the direct coefficient of trust on activism. Although one would expect a higher degree of activism there are reasons why trustful people are less likely to be activists. Generalized trust in people and politics makes protest less necessary because when there is sufficient trust in what people and politicians do, no pressure via activism needs to be built up; hence less respective action is taken.

Social positions (NZ 6): Only five people indicate having the lowest formal education so this group is combined with the 204 persons that state that they have no formal education. There is no one who has never had a paid job; therefore, the reference category is changed to those who are currently unemployed. The 'no answer' category in the question about the community size is also dropped because everyone provided a substantive answer. As the question about the number of toddlers in the household was not asked in New Zealand, no estimates can be calculated.

Starting with income there seems to be ambiguity in New Zealand. While a higher income leads to lower cognitive environmental concern it increases its evaluation and the acceptance of laws. Maybe this is an indication that economically less well-endowed people are more often or more intensely exposed to environmental pollution and thus perceive it more strongly. On the other hand the higher concern in the evaluative dimension can be a function of the 'norms' that come with higher social status; the effects are not very strong, though. This is supported by a higher acceptance of laws connected with higher income. Total effects are only significant on everyday action and laws. As in the Czech Republic, the main effect of income on activism is not significant but the significant coefficient of the squared income indicates that at a certain level of income activism decreases. No other non-linearity effects are present.

The effect of education is almost linear with regard to the evaluative component resulting in the highest concern of those with a university degree. University graduates and those with a higher secondary education also perceive environmental threats as more dangerous. The three best educated groups likewise show a higher degree of activism than the two lower educated groups. These coefficients are almost equal in magnitude and do not differ significantly from each other which in turn indicates that within these three groups there are no differences. The total effects of all groups on all dimensions of action are significant with the best educated registering the highest values.

People's cognitive environmental concern also increases if they are currently employed, while the willingness to pay decreases to a degree almost equal to the increase in the cognitive dimension. Many of the people with a job (especially in manufacturing and agriculture[18]) might be in a position in which they are exposed to more pollution and thus are more concerned and yet less willing to do something about this. With regard to action, only the acceptance of laws is higher among people with a job (total effect).

People living on a farm perceive threats as marginally less dangerous but act a little more environmentally benign (everyday action and activism). The first, albeit small, coefficient may be explained by the same mechanism as in other countries: pollution in the countryside is lower, and likewise its perception. When living at a distance from larger 'settlements', a better management of water, energy or supplies might be necessary, therefore, saving water and energy in particular may be more pronounced on farms. The farmers' greater engagement in activism can be explained by the fact that they might be most dependent, compared to other groups, on a healthy environment and thus protest somewhat more often. The second community type in which people differ from urban inhabitants is the suburban area. They perceive threats to be more dangerous, which translates into a total but very small effect on the three kinds of action.

18 Both are important pillars of New Zealand's economy (New Zealand Government 2013, p. 14).

Table NZ 7: Full model New Zealand

	Cognitive	Evaluative	Conative	Action	Action (total)	Laws	Laws (total)	Activism	Activism (total)
Gender	0.182*			0.134*	0.217*	0.138*			−0.081*
	(0.086/0.271)			(0.049/0.232)	(0.126/0.319)	(0.016/0.257)			(−0.132/−0.040)
Born before 1934 (ref.)									
1934–1943									
1944–1953	0.170*				0.076*		−0.097*		−0.078*
	(0.035/0.307)				(0.011/0.146)		(−0.172/−0.024)		(−0.147/−0.018)
1954–1963	0.188*				0.084*		−0.110*		−0.087*
	(0.052/0.318)				(0.018/0.147)		(−0.181/−0.036)		(−0.149/−0.026)
1964–1973	0.245*				0.112*	0.247*			−0.107*
	(0.107/0.377)				(0.045/0.182)	(0.061/0.423)			(−0.174/−0.043)
1974–1983	0.285*	0.163*			0.140*	0.385*			−0.201*
	(0.129/0.434)	(0.015/0.318)			(0.063/0.217)	(0.196/0.555)			(−0.310/−0.102)
Born after 1983				−0.173*	−0.173*	0.373*	0.373*		
				(−0.313/−0.013)	(−0.312/−0.013)	(0.173/0.571)	(0.173/0.571)		
Post-materialists (ref.)									
Middle	−0.216*		−0.244*		−0.184*			0.413*	0.569*
	(−0.411/−0.036)		(−0.459/−0.097)		(−0.321/−0.080)			(0.136/0.723)	(0.265/0.865)
Materialists	−0.215*		−0.329*		−0.214*	−0.236*		0.542*	0.718*
	(−0.429/−0.006)		(−0.531/−0.172)		(−0.360/−0.101)	(−0.509/−0.007)		(0.227/0.854)	(0.390/10.033)
No answer (ref.)									
Far left	0.266*	0.642*	0.173*	0.261*	0.493*		−0.429*	−0.676*	−10.116*
	(0.105/0.483)	(0.420/0.862)	(0.001/0.356)	(0.083/0.439)	(0.320/0.667)		(−0.583/−0.269)	(−10.011/−0.359)	(−10.458/−0.780)
Left		0.100*			0.008*		−0.051*		−0.189*
		(−0.001/0.213)			(0.000/0.023)		(−0.109/−0.001)		(−0.357/−0.030)
Central									

Table NZ 7: Continued

	Cognitive	Evaluative	Conative	Action	Action (total)	Laws	Laws (total)	Activism	Activism (total)
Right/Conservative	−0.272*				−0.214*	−0.153*			00.116*
	(−0.379/−0.163)				(−0.314/−0.113)	(−0.277/−0.032)			(00.066/00.178)
Far right	−0.494*	−0.317*			−0.250*		0.393*		0.365*
	(−0.769/−0.237)	(−0.5615–0.053)			(−0.398/−0.128)		(0.218/0.589)		(0.202/0.549)
Non-voter									
Trust	−0.323*	00.337*	0.176*			−0.246*	−00.267*	0.379*	
	(−0.567/−0.071)	(00.136/00.593)	(−0.003/0.335)			(−0.560/−0.018)	(−00.577/00.004)	(0.022/0.703)	
Equivalence income		0.099*					−0.081*		
		(0.052/0.148)					(−0.145/−0.022)		
Equivalence income²								−0.046*	−0.046*
								(−0.083/−0.002)	(−0.083/−0.002)
Lowest/No formal education (ref.)									
Intermediate		0.270*			0.018*		−0.148*	−0.249*	−0.376*
		(0.115/0.459)			(0.003/0.043)		(−0.255/−0.075)	(−0.482/−0.044)	(−0.634/−0.142)
Higher secondary	0.128*	0.323*			0.080*	−0.200*	−0.442*	−0.252*	−0.460*
	(0.011/0.245)	(0.188/0.469)			(0.024/0.140)	(−0.616/−0.022)	(−0.655/−0.240)	(−0.432/−0.058)	(−0.665/−0.254)
University incomplete		0.279*			0.018*	−0.161*	−0.332*	−0.242*	−0.385*
		(0.188/0.420)			(0.002/0.042)	(−0.319/−0.002)	(−0.493/−0.162)	(−0.418/−0.080)	(−0.555/−0.217)
University	0.158*	0.550*			0.106*	−0.262*	−0.651*		−0.543*
	(0.029/0.285)	(0.379/0.711)			(0.039/0.179)	(−0.443/−0.054)	(−0.816/−0.453)		(−0.747/−0.338)
Never had paid work	n.a.	n.a.	n.a.	n.a.	n.a.	n.a.	n.a.	n.a.	n.a.
No paid work (ref.)									

Table NZ 7: Continued

	Cognitive	Evaluative	Conative	Action	Action (total)	Laws	Laws (total)	Activism	Activism (total)
In paid work			−0.117*		−0.048*		−0.048*		
			(−0.188/−0.053)		(−0.082/−0.021)		(−0.089/−0.013)		
NA work	0.167*				0.076*		−0.088*		−0.077*
	(0.004/0.340)				(0.001/0.159)		(−0.180/0.000)		(−0.162/−0.004)
School children in household						−0.078*			
						(−0.141/−0.012)			
Toddlers in household	n.a.	n.a.	n.a.	n.a.	n.a.	n.a.	n.a.	n.a.	n.a.
Big city (ref.)									
Suburbs	0.112*				0.051*		−0.060*		−0.050*
	(0.014/0.224)				(0.006/0.103)		(−0.119/−0.007)		(−0.102/−0.005)
Town									
Village									
Farm					0.177*			−0.332*	−0.378*
					(0.047/0.310)			(−0.569/−0.116)	(−0.617/−0.161)
Cognitive		0.425*	0.175*	0.362*	0.457*	−0.329*	−0.518*	−0.215*	−0.443*
		(0.341/0.516)	(0.100/0.251)	(0.271/0.459)	(0.364/0.553)	(−0.437/−0.225)	(−0.622/−0.426)	(−0.342/−0.100)	(−0.552/−0.338)
Evaluative			0.223*		0.078*	−0.444*	−0.444*	−0.375*	−0.430*
			(0.134/0.313)		(0.040/0.127)	(−0.575/−0.327)	(−0.575/−0.327)	(−0.530/−0.232)	(−0.582/−0.285)
Conative				0.362*	0.351*			−0.252*	−0.252*
				(0.271/0.459)	(0.241/0.447)			(−0.440/−0.070)	(−0.440/−0.070)
R^2	0.126	0.368	0.281	0.370	0.370	0.255	0.255	0.256	0.253
Model Fit	CFI=0.958; RMSEA=0.033; SRMR=0.0								

Notes: *p<0.05; unstandardized coefficients; bootstrapped confidence intervals below coefficient in parentheses; only significant coefficients are shown; negative coefficients on laws and activism indicate more acceptance of laws or more activism.

N=1172. Source: ISSP 2010, own calculations.

The full model (Table NZ 7): Besides the general slight drop in coefficient sizes due to the inclusion of further predictors some substantive changes emerge in the full model of New Zealand. When including the measure of general values, the negative effects of the birth cohorts on the conative dimension disappear. Especially post-materialist values seem to 'steal' the impact of the birth cohorts. Due to this loss of significance, the total effects climb up and some that were not significant in Table NZ 4 become significant in Table NZ 7. These 'new' effects are not very large. In the same momentum, the only significant cohort effect (1944–1953) on activism disappears in the full model once controlled for general values.

The once positive coefficient of the right/conservatives on the evaluation of environmental issues disappears. This entails a loss of significance of the total effect on laws but leads to an increased total negative effect on everyday action. The total effect on activism now becomes significant indicating less activism among right/conservatives.

In New Zealand most of the direct effects encountered in Table NZ 7 take place on the attitudinal level rather than on everyday action or activism. For the acceptance of laws some more direct effects were found. In total cohorts/gender, general values and social position contribute only marginally to the explanation of action. In contrast, as for all three activities environmental attitudes are quite substantial predictors, this study indicates where decision-makers could start to promote environmental concern; and in the end increase environmental action. Coming back to Europe, Norway, a social democratic welfare state will be visited next. A country where on the one hand people are least engaged in everyday action (Table 5.1) but on the other hand rank the environment to be a more important issue than the economy (Table NO 1).

6.4 Norway

GPD per capita (2010): 85,443 $

Norway seems to be a special case concerning the environment. Grendstad and colleagues (2006) characterized Norway's "Unique Environmentalism". That the environment and environmentalism is a special case in Norway is particularly traced back to its political culture. Norway has a distinctive state structure and civil society that differentiate Norway from most other nations (Grendstad *et al.* 2006, p. 1). It is "an inclusive polity in a state-friendly society" (Grendstad *et al.* 2006, p. 15). In this society the state plays an active and interventionist role and can count on wide support and legitimacy from the public. The 'strong' state is complemented by a high level of citizen control and the importance of voluntary organizations and movements in the nation-building process (Grendstad *et al.*

2006, pp. 16–17). The second unique characteristic is the "local community perspective" (Grendstad *et al.* 2006, p. 21). Norway is a country with vast wilderness and sparse population and thus strong ties to nature have developed. In the same vein a culture of the "ideal of the local self-reliant community" (Grendstad *et al.* 2006, p. 23) evolved. The principle of the local community perspective is the "protection of humans in nature" (Grendstad *et al.* 2006, p. 23). As Table NO 1 reveals, in contrast to most nations and the average, Norwegians consider health care instead of the economy the most important issue. After health care education is second most important topic. Before the economy is encountered in Table NO 1 the next issue is the environment on rank three. This finding reflects the special case of Norway's environmentalism described above. The environment's position is notable not only because it is an exception in contrast to the other nations. Norwegian people also perceive environmental problems least severe and act least environmentally friendly in their day to day activities, while in the acceptance of laws and activism Norway is found on middle to lower middle positions (Table 5.1). Maybe due to the relatively high salience of the environment in the Norwegian society and respective policies the environmental quality is relatively good and consequently environmental threats are perceived less intensely and less action is required. Additionally, the active interventionist state mentioned above may play a more important role in dealing with environmental issues than people as individual actors do. The high concern in ranking items and medium to low concern in rating items underline, that there might be a substantial difference whether rating or ranking items are used in a survey.

Nawrotzki (2012) applies the same measure of the willingness to pay scale as in the present study. The author analyzes the ISSP 2000 and finds Norway on rank seven with the same mean as Germany West. In the ISSP 2010 Norway ranks eighth with a slightly lower (but not significantly different) willingness to pay than Germany West. Obviously, not so much changed in the past ten years. Franzen and Meyer (2010) report descriptive statistics for their very broad index of environmental concern and find Norway on an upper middle position with the ISSP 2000 data. In the ISSP 1993 Norway holds the fourth place (Diekmann and Franzen 1999). Inglehart (1995) also discovers Norway on a fifth place concerning its willingness to pay and the seventh last place (both out of 39) concerning its unwillingness to pay. This is a contrast, for instance, to the findings of the Czech Republic or Spain which rather hold high positions in the Inglehart study (WVS) and lower positions in the studies using the ISSP. If only looking at the rating scale applied in the cited studies Norway would seem to be an averagely concerned country. This can be delusive when recalling the high importance of the environment in relation to other issues in Table NO 1.

Table NO 1: Most important issue in Norway

	Most important	Second most important	Score
Health care	538	307	691.5
Education	234	289	378.5
The environment	207	212	313
The economy	124	150	199
Crime	68	164	150
Immigration	88	117	146.5
Poverty	35	53	61.5
None of these	20	13	26.5
Terrorism	11	25	23.5

Notes: Source: ISSP 2010, own calculations.

Table NO 2 shows that the three most important environmental issues in Norway are those that form the top three in the six nation's average (Table 6.2) although in Norway climate change ranks first and air pollution second. Most surprising is the finding that the 'none of these' category is found on the fourth place. This is especially due to the 229 votes that stated none of these issues is the most important for the family. This might again be an indication that, at least on a micro level, people in Norway are not that strongly burdened with environmental problems. When ignoring this category the three least important problems are the same three as on average (Table 6.2) only that the order is altered: nuclear waste, genetically modified foods and water shortages.

Of the six investigated country samples, Norwegian respondents are the strongest promoters of renewable energies: 84.5% state that solar, wind or water power is the most important source to meet the countries future energy need (Table NO 3). This number is slightly above the number of Germany West (81%) and sets itself apart from the average of 68.5% (Table 6.3). With 4.9% only Germany West considers nuclear energy less important than people in Norway where 5.9% think this would be the best source to meet future energy demands. This also reflects in real numbers: 100% of the Norwegian Energy production originates from renewable resources. About 95% stem from hydropower, ca. 4% from thermal power and the last 1% from wind power (Norwegian Water Resources and Energy Directorate 2011). Obviously, there exist neither nuclear power plants nor coal fired power plants on Norwegian ground. However, the energy mix to supply Norway with energy looks different due to ex- and imports of energy. About 31.4% of its total primary energy is supplied by solar, wind and water power while fossil fuels still have a share of 61.3%.

Table NO 2: Most important environmental issues in Norway

	For country	For family	Score
Climate change	351	179	265
Air pollution	246	283	265
Using up our natural resources	251	101	176
None of these	22	229	126
Domestic waste disposal	53	135	94
Water pollution	115	61	88
Chemicals and pesticides	80	77	79
Nuclear waste	95	27	61
Genetically modified foods	47	66	57
Water shortage	46	58	52

Notes: Source: ISSP 2010, own calculations.

Table NO 3: Energy supply in Norway

	Best source for future energy supply		Real total primary energy supply (TPES) 2010
	%	N	
Solar, wind or water power	84.5%	1108	31.4%
Nuclear power	5.9%	77	0.0%
Coal, oil and natural gas	4.7%	62	61.3%
Fuels made from crops	4.3%	57	5.1%
None of them	0.6%	8	2.2%
Total	100%	1312	100%

Notes: Source: ISSP 2010, own calculations. International Energy Agency (2012).

Multivariate analysis

The model fit of the basic model in Norway is only acceptable (CFI=0.922; RMSEA=0.071; SRMR=0.058). Cognitive → affective = 0.632 (0.529/0.734); cognitive → conative=0.103 (0.004/0.195); affective → conative=0.635 (0.537/0.739).

Table NO 4: Cohorts, gender and environmentalism in Norway

	Cognitive	Evaluative	Conative	Action	Action (total)	Laws	Laws (total)	Activism	Activism (total)
Male (ref.)									
Female	0.318*	0.088*		0.100*	0.242*	0.154*			−0.143*
	(0.238/0.403)	(0.011/0.161)		(0.015/0.184)	(0.154/0.324)	(0.059/0.256)			(−0.187/−0.097)
Born before 1944 (ref.)									
1934–1943	n.a.	n.a.	n.a.	n.a.	n.a.	n.a.	n.a.	n.a.	n.a.
1944–1953		0.177*	−0.134*						
		(0.007/0.227)	(−0.264/−0.008)						
1954–1963		0.285*	−0.181*				−0.157*		
		(0.176/0.397)	(−0.316/−0.051)				(−0.241/−0.076)		
1964–1973	0.228*	0.258*	−0.396*				−0.250*		
	(0.118/0.338)	(0.140/0.369)	(−0.527/−0.257)				(−0.359/−0.146)		
1974–1983	0.343*	0.282*	−0.266*	−0.145*		0.141*	−0.208*		−0.152*
	(0.210/0.465)	(0.161/0.415)	(−0.407/−0.121)	(−0.278/−0.025)		(−0.001/0.280)	(−0.366/−0.044)		(−0.236/−0.073)
Born after 1983	0.208*		−0.322*	−0.270*	0.271*	0.267*	0.180*	−0.201*	−0.186*
	(0.077/0.342)		(−0.464/−0.182)	(−0.408/−0.132)	(−0.427/−0.124)	(0.121/0.414)	(0.015/0.351)	(−0.344/−0.054)	(−0.333/−0.021)
R^2	0.091	0.422	0.455	0.257	0.257	0.279	0.279	0.144	0.144
Model fit	CFI=0.927; RMSEA=0.056; SRMR=0.045								

Notes: *p<0.05; unstandardized coefficients; bootstrapped confidence intervals below coefficient in parentheses; only significant coefficients are shown; negative coefficients on laws and activism indicate more acceptance of laws or more activism. n.a.: no coefficient was calculated because no respondent chose the category. N=1382.

Source: ISSP 2010, own calculations.

An investigation of the factor loadings does not reveal any major misspecifications (all amount to >0.5). The modification indices, however, indicate various relationships between the observed variables that go beyond the relations of the dimensions they belong to, despite the relatively strong paths from the cognitive dimension to the evaluative dimension (b=0.623) and from the evaluative dimension to the conative factor (b=0.635). Many minor cross-loadings are implied by the modification indices. Most notable is an implied loading of the evaluative dimension of concern and item Q14e (severity of climate change; see 4.2.3). Estimating the cross-loading or deleting Q14e improves the fit but only negligibly changes the path coefficients. As with Germany East or New Zealand, for comparability reasons and due to the fact that the fit is still acceptable, the basic model will be retained.

Cohorts and gender (NO 4): The gender-environment relation (Table NO 4) resembles that of New Zealand: women are more environmentally friendly in Norway (cognitive and action) but to a greater extent than in New Zealand. In Norway gender effects are quite straightforward and in accordance with most findings of previous research: women perceive environmental threats as more severe, evaluate environmental issues more pro-environmentally, and moreover they are marginally more active in their day-to-day routine. This results in the total significant effects on action and activism. At the same time they support laws less strongly than men; still, due to their higher level of environmental concern, this effect is canceled out and thus the total effect is not significant.

The Norwegian sample is slightly younger than the other samples, featuring only 13 respondents born before 1934. Hence, the reference category in Norway includes all those born before 1944. Even though this lumps the war and prewar generation together, the results point in the same direction, although the estimates are more precisely portrayed by smaller confidence intervals in comparison to the analysis where the two oldest cohorts are separated (results not shown). Significant effects of birth cohort are found on the cognitive, evaluative and conative dimension as well as on the three types of action. These effects are contradictory. While the three cohorts from 1954 to 1983 are more concerned in the first two dimensions they are less willing to pay. The 1974–1983 cohort is also less active and accepts laws less strongly but scores higher on activism (total effect). The youngest generation in the sample also has to deal with a similar ambiguity: they perceive environmental threats to be more dangerous but they are less willing to do something about it and also are less active in their daily routine. They prefer to show their higher environmental concern through activism.

Table NO 5: General values and environmentalism in Norway

	Cognitive	Evaluative	Conative	Action	Action (total)	Laws	Laws (total)	Activism	Activism (total)
Post-materialists (ref.)									
Mixed type		−0.160*	−0.136*		−0.060*		0.098*	0.289*	0.370*
		(−0.289/−0.030)	(−0.260/−0.009)		(−0.106/−0.097)		(0.033/0.169)	(0.130/0.454)	(0.207/0.534)
Materialists	−0.135*	−0.378*	−0.199*		−0.160*		0.291*	0.286*	0.493*
	(−0.236/−0.032)	(−0.528/−0.234)	(−0.341/−0.052)		(−0.236/−0.022)		(0.198/0.391)	(0.096/0.469)	(0.311/0.667)
No answer (ref.)									
Far left	0.549*	0.345*			0.260*		−0.464*	−1.045*	−1.326*
	(0.164/0.964)	(−0.002/0.691)			(0.112/0.447)		(−0.755/−0.211)	(−1.485/−0.610)	(−1.831/−0.880)
Central/left		0.197*			0.029*		−0.106*		−0.062*
		(0.108/0.284)			(0.014/0.048)		(−0.169/−0.056)		(−0.110/−0.029)
Central		0.179*	0.188*		0.076*		−0.116*		−0.100*
		(0.072/0.296)	(0.096/0.297)		(0.041/0.119)		(−0.183/−0.057)		(−0.151/−0.060)
Right/conservative	−0.374*		−0.099*		−0.170*		0.210*		0.144*
	(−0.468/−0.281)		(0.184/−0.012)		(−0.222/−0.123)		(0.157/0.276)		(0.101/0.194)
Far right	n.a.	n.a.	n.a.	n.a.	n.a.	n.a.	n.a.	n.a.	n.a.
Non voter								−0.475*	−0.526*
								(−0.686/−0.251)	(−0.759/−0.299)
Trust		0.891*	0.463*		0.253*	−0.609*	−1.124*		−0.383*
		(0.688/1.106)	(0.249/0.691)		(0.158/0.355)	(−0.921/−0.328)	(−1.399/−0.872)		(−0.507/−0.268)
R^2	0.075	0.519	0.468	0.234	0.234	0.276	0.276	0.173	0.173
Model fit	CFI=0.940; RMSEA=0.050; SRMR=0.043								

Notes: *p<0.05; unstandardized coefficients; bootstrapped confidence intervals below coefficient in parentheses; only significant coefficients are shown; negative coefficients on laws and activism indicate more acceptance of laws or more activism. n.a.: no coefficient was calculated because no respondent chose the category. N=1382.

Source: ISSP 2010, own calculations.

The discrepancy between aspiration and reality seems well pronounced in Norway with regard to cohort effects. When recalling that in Norway the environment was ranked third in comparison to other issues (Table NO 1) but the amount of everyday action was lowest (Table 5.1), it is reasonable to assume that the way the Norwegians of different birth cohorts contribute to environmental protection takes place in other activities. Younger ones more often engage in activism while older people instead act environmentally friendly in their day-to-day routine and accept laws more strongly. Cohort-specific attention shifting is at work.

General values (NO 5): The importance of postmaterialism reveals itself in negative and significant coefficients of both categories on the intention to do something and the evaluation of environmental issues. A negative effect of materialists on the cognition of environmental threats is likewise present; everyday action as well as the acceptance of laws are not affected directly, while activism is. The extent of the effects of mixed types and materialists is somewhat greater than in Germany East and the Czech Republic; with regard to the total effects on the three types of action, it is comparable to those of Germany West.

The single difference from the other samples in the trust-effects is the non-significant coefficient on the cognitive dimension. The evaluative and conative dimensions are affected positively. Due to the absence of a negative effect on the cognitive dimension, the total effects are not reduced and are therefore significant in all three types of action. The total effect on the acceptance of laws is boosted by an additional direct effect of trust.

In Norway none of the respondents are classified as far right; consequently, this category is left out of the analysis. The role that political position plays in Norway varies depending on the aspect of environmental concern but shows signs as expected. Far-left voters perceive environmental threats to a greater degree and register higher scores in the evaluative dimension than most other groups; furthermore, they are more often engaged in activism. Interestingly, non-voters are also likely to be activists. In Norway, non-voting does not ultimately indicate political disinterest, rather, political interest is expressed via activism. More notable is the negative coefficient of conservatives who quite clearly judge environmental problems to be less dangerous. The lefts and central voters are also more evaluatively concerned. The central voters, moreover, are most willing to do something. Except for the far-left voters and the non-voters' direct effects on activism, none of the political positions affects any kind of action directly.

Table NO 6: Social positions and environmentalism in Norway

	Cognitive	Evaluative	Conative	Action	Action (total)	Laws	Laws (total)	Activism	Activism (total)
Equivalence income		–0.059*			–0.016*	0.116*	0.149*		0.020*
		(–0.097/–0.023)			(–0.031/–0.006)	(0.067/0.165)	(0.097/0.199)		(0.008/0.037)
Equivalence income²									
Lowest form. education (ref.)									
Intermediate					0.005*		–0.046*	–0.236*	–0.242*
					(0.001/0.014)		(–0.093/–0.014)	(–0.458/–0.009)	(–0.464/–0.014)
Secondary					0.004*		–0.040*	–0.263*	–0.269*
					(0.001/0.012)		(–0.082/–0.009)	(–0.465/–0.037)	(–0.473/–0.042)
University incomplete					0.007*	–0.156*	–0.218*	–0.306*	–0.314*
					(0.002/0.017)	(–0.297/–0.012)	(–0.359/–0.069)	(–0.529/–0.076)	(–0.536/–0.087)
University complete	0.148*	0.301*		–0.144*		–0.123*	–0.431*	–0.290*	–0.452*
	(0.061/0.247)	(0.222/0.381)		(–0.238/–0.053)		(–0.236/–0.003)	(–0.558/–0.314)	(–0.511/–0.072)	(–0.677/–0.242)
No answer					–0.015*		0.139*		0.019*
					(–0.037/–0.005)		(0.055/0.255)		(0.005/0.043)
Never had paid work (ref.)									
No paid work					0.019*		–0.182*		–0.024*
					(0.007/0.044)		(–0.276/–0.106)		(–0.052/–0.010)
In paid work		0.086*	–0.129*	–0.104*			–0.247*		
		(0.005/0.173)	(–0.215/–0.048)	(–0.200/–0.009)			(–0.362/–0.155)		
No answer					0.110*				–0.140*
					(0.007/0.245)				(–0.329/–0.017)
School children in household			–0.043*		–0.010*	–0.064*	–0.056*		0.012*
			(–0.008/–0.005)		(–0.023/–0.002)	(–0.188/–0.011)	(–0.113/–0.005)		(0.000/0.026)

Table NO 6: Continued

	Cognitive	Evaluative	Conative	Action	Action (total)	Laws	Laws (total)	Activism	Activism (total)
Toddlers in household	0.097*				0.106*		−0.050*		
	(0.028/0.171)				(0.026/0.186)		(−0.090/−0.012)		
Big city (ref.)									
Suburbs	−0.174*				−0.069*		0.094*		0.063*
	(−0.326/−0.035)				(−0.134/−0.014)		(0.017/0.178)		(0.012/0.123)
Town	−0.124*	−0.094*			−0.074*		0.114*		0.077*
	(−0.240/−0.005)	(−0.183/−0.012)			(−0.027/−0.128)		(0.042/0.187)		(0.029/0.131)
Village	−0.138*						0.073*		0.050*
	(−0.270/−0.024)						(0.009/0.143)		(0.008/0.099)
Farm	−0.262*				−0.103*	0.131*	0.268*		0.094*
	(−0.387/−0.141)				(−0.053/−0.053)	(0.005/0.260)	(0.124/0.412)		(0.045/0.146)
No answer					−0.034*				0.043*
					(−0.068/−0.013)				(0.017/0.083)
R^2	0.039	0.447	0.436	0.252	0.252	0.288	0.288	0.146	0.146
Model fit	CFI=0.958; RMSEA=0.039; SRMR=0.034								

Notes: *$p<0.05$; unstandardized coefficients; bootstrapped confidence intervals below coefficient in parentheses; only significant coefficients are shown; negative coefficients on laws and activism indicate more acceptance of laws or more activism. N=1382.

Source: ISSP 2010, own calculations.

Their impact is only mediated via environmental concern, mirrored by the significant total effects. Once more evidence is found that political positions and the way of impacting environmentalism varies between countries or different aspects of one issue; with people more on the left wing being most environmentally concerned in Norway.

Social positions (NO 6): In the Norwegian sample only five persons stated that they do not have any formal education. These are combined with the 'lowest formal education' category which is the reference category. In all other variables a sufficient number of persons fall into each category so none of them has to be excluded from the analysis.

As Table NO 6 reveals, social positions neither contribute to environmental concern nor to any form of action excessively. A minor negative impact comes from the equivalence income on the evaluative concern while the other dimensions are unaffected. The income effect is mediated into lower everyday action, lower support for laws and less activism. Additionally, a direct effect emerges from a person's household income that lowers his or her support for laws. No significant coefficients for the squared income are present.

Surprisingly, education also only plays a minor role compared to in the Czech Republic, Germany or New Zealand: first, people with a university degree hold higher values on the evaluative dimension but this dimension is the weakest predictor of action among the three dimensions of concern. They also perceive threats to be more dangerous, yet despite this a university degree reduces everyday action. They shift their attention to activities that are more symbolic, i.e. they exhibit higher scores (both directly and indirectly) in law-acceptance and activism. A solid support for laws is also present among those with an incomplete university degree. All educational groups are more often engaged in activism than those with the lowest education. Even though the coefficients are quite similar, the tendency holds that those of the groups of higher education are somewhat larger. They are not significantly different, though.[19]

Employed people are more concerned in the evaluative dimension but less willing to pay and likewise are less active. However, they support laws more strongly when looking at the total effects. These results show the problems that a social position might entail. A lack of time can explain the lower willingness to pay and everyday action as well as the higher support for laws (recall that instead of action this is non-action). At the same time, the more stable life situation

19 Critical ratios computed by AMOS in pairwise parameter comparison vary between −0.830 and 0.203. The critical ratio is comparable to a z-statistic.

allows people to evaluate other problems more concerned. Mediated via lower income, those without a job at present, show higher values in action, support laws more strongly and are more often involved in activism.

Another small effect stems from the number of toddlers in the household. With a higher number of toddlers at home environmental threats are perceived as somewhat more dangerous. In contrast, having school children at home decreases the willingness to pay but increases the support for laws. The total effects of children in the household are inconsistent: while school children minimally reduce everyday action, their presence increases the acceptance of laws but lowers activism. Toddlers, in contrary, only lead to higher scores in action and a greater acceptance of laws.

The next group of paths investigated is the one comprising those originating from the community size. There is a slightly decrease in cognitive environmental concern with decreasing community size. This accompanies with the assumption that generally the environmental quality rises as the community size decreases and thus environmental threats are perceived less strongly; the coefficients are small, though. Additionally, people from town-sized communities are less concerned by environmental issues. On farms or on country side people reject laws rather than accepting them. It is easy to imagine that environmental legislation that is meant, for instance, to reduce the amount of pesticides used will be met with a farmer's objection, hence the generally lower support for laws. In total, the effects of community size on concern mounts up to less everyday action, less support for laws and less activism in comparison to people from big cities.

The contribution of social positions to the explanation of environmentalism in Norway is quite small. On the one hand some part of environmentalism already has diffused into society and therefore is independent of the social positions people hold. On the other hand, when comparing with the results of the general values in Table NO 5, the part of environmentalism that has not yet diffused into society still tends to be subject to people's general values than to positional opportunities or constraints. How values, position as well as age and gender influence environmentalism conjointly will be investigated in the next step, in the full model.

Full model (Table NO 7): In general, most findings stay as they are in the separate analyses. Slight decreases in the effect size appear and especially some of the smaller coefficients that already were at the limit of practical relevance lose significance (e.g. the small effects of gender on the evaluative dimension or of the materialists' on the cognitive component). The most notable variations, i.e. large drops in size or significance as well as coefficients that only became significant in the full model, are now alluded to.

Table NO 7: Full model Norway

	Cognitive	Evaluative	Conative	Action	Action (total)	Laws	Laws (total)	Activism	Activism (total)
Gender	0.293*			0.100*	0.217*	0.137*			−0.086*
	(0.209/0.370)			(0.011/0.177)	(0.136/0.306)	(0.044/0.231)			(−0.123/−0.057)
Born before 1944 (ref.)									
1944–1953	n.a.	n.a.	n.a.	n.a.	n.a.	n.a.	n.a.	n.a.	n.a.
1954–1963		0.142*	−0.153*				−0.071*		
		(0.049/0.236)	(−0.283/−0.048)				(−0.135/−0.016)		
1964–1973	0.196*	0.156*	−0.355*				−0.159*		
	(0.090/0.315)	(0.057/0.259)	(−0.477/−0.231)				(−0.246/−0.072)		
1974–1983	0.273*	0.178*	−0.222*	−0.153*		0.159*			−0.077*
	(0.142/0.394)	(0.060/0.300)	(−0.363/−0.092)	(−0.286/−0.027)		(0.032/0.276)			(−0.147/−0.012)
Born after 1983	0.151*		−0.272*	−0.254*	−0.262*	0.204*		−0.248*	−0.222*
	(0.010/0.277)		(−0.403/−0.132)	(−0.403/−0.126)	(−0.426/−0.129)	(0.044/0.349)		(−0.399/−0.095)	(−0.387/−0.050)
Post-materialists (ref.)									
Middle		−0.177*		−0.166*	−0.191*		0.077*	0.297*	0.346*
		(−0.278/−0.066)		(−0.306/−0.021)	(−0.332/−0.044)		(0.031/0.134)	(0.111/0.501)	(0.155/0.560)
Materialists		−0.364*	−0.091*	−0.152*	−0.227*		0.174*	0.300*	0.425*
		(−0.498/−0.247)	(−0.180/0.001)	(−0.335/0.000)	(−0.402/−0.064)		(0.112/0.252)	(0.108/0.508)	(0.238/0.641)
No answer (ref.)									
Far left	0.516*				0.208*		−0.276*	−10.127*	−10.278*
	(0.457/0.659)				(0.046/0.420)		(−0.530/−0.051)	(−10.834/−0.396)	(−20.047/−0.483)
Left		0.132*			0.020*		−0.081*	−0.100*	−0.139*
		(0.045/0.216)			(0.008/0.037)		(−0.126/−0.039)	(−0.191/−0.006)	(−0.232/−0.040)

Table NO 7: Continued

	Cognitive	Evaluative	Conative	Action	Action (total)	Laws	Laws (total)	Activism	Activism (total)
Central		0.120*	0.168*		0.059*		–0.078*		–0.078*
		(0.015/0.217)	(0.065/0.283)		(0.027/0.099)		(–0.133/–0.031)		(–0.128/–0.039)
Right/ Conservative	–0.355*		–0.119*		–0.172*		0.196*		0.137*
	(–0.455/–0.262)		(–0.197/–0.041)		(-–0.229/–0.128)		(0.136/0.267)		(0.097/0.188)
Far right									
Non-voter								–0.552*	–0.552*
								(–0.906/–0.251)	(–0.906/–0.251)
Trust		0.794*	0.406*		0.218*	–0.561*	–1.012*		–0.332*
		(0.556/10.012)	(0.172/0.669)		(0.125/0.320)	(–0.889/–0.258)	(–1.308/–0.732)		(–0.475/–0.220)
Equivalence income		–0.041*			–0.006*	0.112*	0.131*		0.011*
		(–0.080/–0.004)			(–0.013/–0.001)	(0.051/0.170)	(0.074/0.189)		(0.002/0.026)
Equivalence income²									
Lowest formal education (ref.)									
Intermediate								–0.255*	–0.299*
								(–0.421/–0.084)	(–0.393/–0.056)
Higher secondary					–0.049*			–0.281*	–0.248*
					(–0.087/–0.009)			(–0.445/–0.115)	(–0.411/–0.085)
University incomplete					0.003*	–0.160*	–0.222*	–0.320*	–0.326*
					(0.001/0.009)	(–0.307/–0.022)	(–0.386/–0.083)	(–0.495/–0.133)	(–0.501/–0.141)

Table NO 7: Continued

	Cognitive	Evaluative	Conative	Action	Action (total)	Laws	Laws (total)	Activism	Activism (total)
University		0.215*		−0.125*	−0.091*	−0.144*	−0.305*	−0.304*	−0.370*
		(0.145/0.294)		(−0.219/−0.039)	(−0.186/−0.004)	(−0.247/−0.036)	(−0.417/−0.186)	(−0.478/−0.138)	(−0.543/−0.201)
Never had paid work (ref.)									
No paid work					0.006*		−0.125*	−0.456*	−0.467*
					(0.001/0.016)		(−0.240/−0.050)	(−0.766/−0.234)	(−0.776/−0.244)
In paid work							−0.147*	−0.415*	−0.428*
							(−0.272/−0.069)	(−0.706/−0.204)	(−0.713/−0.211)
NA work		0.349*			0.050*		−0.146*	−0.623*	−0.720*
		(0.150/0.554)			(0.020/0.096)		(−0.236/−0.066)	(−10.242/−0.050)	(−10.339/−0.181)
School children in household						−0.054*			
						(−0.106/−0.003)			
Toddlers in household				0.102*	0.092*				0.019*
				(0.016/0.177)	(0.001/0.165)				(0.000/0.047)
Big city (ref.)									
Suburbs	−0.147*				−0.059*		0.085*		0.045*
	(−0.278/0.001)				(−0.177/−0.001)		(0.009/0.166)		(0.004/0.090)
Town	−0.111*	−0.096*			−0.059*		0.103*		0.061*
	(−0.255/0.000)	(−0.178/−0.011)			(−0.105/−0.013)		(0.028/0.177)		(0.018/0.105)
Village	−0.123*						0.069*		0.038*
	(−0.242/−0.001)						(−0.001/0.137)		(0.000/0.076)
Farm	−0.258*				−0.104*		0.149*		0.079* (0.038/0.124)

Table NO 7: Continued

	Cognitive	Evaluative	Conative	Action	Action (total)	Laws	Laws (total)	Activism	Activism (total)
	(−0.376/−0.130)				(−0.163/−0.054)		(0.077/0.226)		
Cognitive		0.559*	0.146*	0.286*	0.402*	−0.289*	−0.557*	−0.110*	−0.301*
		(0.457/0.659)	(0.020/0.246)	(0.194/0.380)	(0.311/0.491)	(−0.404/−0.160)	(−0.649/−0.474)	(−0.223/−0.005)	(−0.380/−0.220)
Evaluative			0.568*		0.143*	−0.373*	−0.450*		−0.277*
			(0.458/0.703)		(0.086/0.213)	(−0.541/−0.211)	(−0.602/−0.316)		(−0.415/−0.157)
Conative				0.252*	0.252*	−0.138*	−0.138*	−0.265*	−0.265*
				(0.149/0.343)	(0.149/0.343)	(−0.248/−0.046)	(−0.248/−0.046)	(−0.382/−0.142)	(−0.382/−0.142)
R^2	0.177	0.563	0.481	0.277	0.277	0.305	0.305	0.183	0.183
Model fit		CFI=0.981; RMSEA=0.027; SRMR=0.022							

Notes: *$p<0.05$; unstandardized coefficients; bootstrapped confidence intervals below coefficient in parentheses; only significant coefficients are shown; negative coefficients on laws and activism indicate more acceptance of laws or more activism. N=1382.

Source: ISSP 2010, own calculations.

Of importance are the now significant direct effects of mixed types and materialists on everyday action which, of course, also increase the respective total effects. The coefficient of the left voters on activism is also significant in the full model.

The loss of significance of a university degree on the perception of threats leads to a now significant (but very small) total negative effect on everyday action.

Most notable are the changed coefficients of the occupational position when including the birth cohorts. The effects on environmental concern disappear: like in the other country samples, those with a job are mostly middle aged, those who never had a paid job are predominantly young and those with no job at the moment are older and often already retired. With regard to activism, under control of the cohorts, quite substantial direct effects emerge. Those with a job and those currently with no job are more likely to be involved in activism than people who had never had a paid job so far. The age structure of the people concealed these substantial coefficients.

In sum, general values in particular should not be disregarded when considering environmental concern in Norway. The effect of concern on action is not the strongest in Norway but it is substantive and thus when trying to promote environmental action people's values and attitudes should not be neglected. This is especially true for the acceptance of laws and activism, where direct effects are present. Likewise, generational differences need to be addressed. Of lesser import are the social positions people occupy. By now leaving the northern latitudes of wealthy and economically stable Norway, this study turns its focus of investigation onto the economically-hit Mediterranean region, to Spain.

6.5 Spain

GDP per capita (2010): $ 29,956

In total, research so far has shown that the Spanish people hold a more or less middle to lower position concerning their environmentalism – depending on the data used. Diekmann and Franzen (1999) find an average degree of environmental concern using ISSP data of 1993. In Franzen and Meyer (2010) or Nawrotzki (2012) in their results based on the ISSP 2000 the Spanish can 'defend' their middle position. In contrast, Inglehart's (1995) analysis of the World Values Survey positions Spain close to the bottom third concerning their willingness to pay higher taxes, in second position (out of 39) regarding people's unwillingness to pay and fifth lowest in an index of three items indicating people's marginal willingness to pay.[20]

20 This index includes the "willingness to pay higher taxes", the "willingness to pay higher prices" and the unwillingness to pay measured in form of an agreement or disagreement

In 1997, Dekker et al. analyzed the 1993 ISSP data and found Spanish respondents in a middle position concerning the perception of threats while at that time they still were among the four (out of 20) countries most willing to pay. In the ISSP 2010 the perception of environmental threats is highest in Spain while it holds the second-last position in the evaluative dimension and is in the bottom quarter concerning its willingness to pay (Table 5.1). Hadler and Haller (2011) investigated pro-environmental behavior with the ISSP 2000 data. Descriptive statistics show that the Spanish respondents hold the tenth-highest scores in private behavior (recycling and forgoing car driving) and are only ranked 16th in public behavior (activism), both out of 22. The results of the present study still rank the Spanish 10th regarding both everyday action and the acceptance of laws (Table 5.1). Likewise, a moderately low frequency of activism exists in Spain, positioning the country 14th in this dimension.

According to World Bank data (2013) the Spanish recession started in 2009 with a decline of the GDP by 3.7% and a further 0.3% drop in 2010. One might suggest that already in 2010 people might be bothered more directly by economic and related problems rather than the environment. Indeed the inspection of Table ES 1 clearly shows there is a discrepancy between economy and the environment. These results cannot tell whether this is due to the economic crisis or not. Besides the category 'none of these' the environment ranks last and the economy occupies first place with a score 8.7 times as high as the environment. This is a strong contrast to the average of the other nations where the environment mostly occupies a middle-ranking position. In part, this 'lack' of environmental concern with regard to two ranking items is also reflected by Spain's low degree of evaluative and conative environmental concern (see Table 5.1).

In contrast to the other countries the issue of water shortage is quite important for Spanish people (Table ES 2). Only climate change and air pollution are more often chosen as most important environmental issue. Spain often has to deal with water shortages and droughts (European Environment Agency 2010) which explains why this particular problem is more pronounced than in the other five countries samples. This inspection shows that not only global environmental problems with broad media coverage are perceived to be important but concrete local environmental problems are on the agenda as well.

---

with the statement "the government has to reduce environmental pollution but it should not cost me any money".

Table ES 1: Most important issue in Spain

	Most important	Second most important	Score
The economy	1091	438	1310
Health care	540	442	761
Education	401	551	676.5
Poverty	144	246	267
Terrorism	88	226	201
Immigration	81	219	190.5
Crime	85	206	188
The environment	77	148	151
None of these	24	13	30.5

Notes: Source: ISSP 2010, own calculations.

Table ES 2: Most important environmental issue in Spain

	For country	For family	Score
Climate change	604	375	490
Air pollution	378	468	423
Water shortage	381	262	322
Using up our natural resources	290	233	262
Water pollution	207	237	222
Chemicals and pesticides	215	206	211
Nuclear waste	200	135	168
Genetically modified foods	83	219	151
Domestic waste disposal	95	205	150
None of these	13	97	55

Notes: Source: ISSP 2010, own calculations.

In contrast to Spain's deviation concerning the most important environmental problems, the issue of energy supply is closer to the average. As for the other five nations Spain's main suppliers of primary energy are fossil fuels followed by nuclear energy (Table ES 3). While wind, solar and water power only mount up to 6.6%, 77.6% of the ISSP, respondents state that these sources are the best to meet future energy demand. As in the six countries' average, nuclear power is the second-most important energy source but a long distance behind the first one. Again there is a distinct discrepancy between the actual situation and the people's preferences.

Turning now to the relationship of environmental attitudes and action in Spain, only 16% of variance in everyday action, 7.6% of the acceptance of laws and 13.3% of activism are attributed to environmental concern (Table 5.2), Spain is ranked rather low down. The weakest predictor of everyday action and the acceptance of laws is the conative dimension of concern while for activism the perception of threats does not play a role at all. For all three aspects of action, the evaluative dimension is relatively important and thus action carries a value-rational element. At least for laws and everyday action there is also an instrumentally rational part, as the effects of the cognitive component show.

Table ES 3: Energy supply in Spain

	Best source for future energy supply		**Real total primary energy supply (TPES) 2010**
	%	N	%
Solar, wind or water power	77.6%	1765	6.6%
Nuclear power	9.4%	213	12.6%
Fuels made from crops	7.3%	167	5.3%
Coal, oil and natural gas	4.8%	110	75.9%
None of them	0.8%	19	-0.6%
Total	100.0%	2274	100.0%

Notes: Source: ISSP 2010, own calculations. International Energy Agency (2012).

Multivariate analysis

Basic model: CFI=0.955; RMSEA=0.049; SRMR=0.037; Cognitive → affective=0.464 (0.394/0.536); cognitive → conative=n.s.; affective → conative=0.414 (0.330/0.503).

Cohorts and gender (Table ES 4): Table ES 4 reveals that women are less willing to pay, are less often involved in activism but are slightly more active in their everyday routine than men. The negative effect on the conative dimension reduces the total effect on action and leads to a small reduction in the support for laws as well as activism.

The general tendency that older cohorts are less environmentally concerned (cognitive and evaluative) as found in other countries (e.g. New Zealand or Norway), is also true for Spain. This is a non-linear relationship; the most concerned generations are the same as in New Zealand: 1964–1973 and 1974–1983.

Table ES 4: Cohorts, gender and environmentalism in Spain

	Cognitive	Evaluative	Conative	Action	Action (total)	Laws	Laws (total)	Activism	Activism (total)
Male (ref.)									
Female			−0.070*	0.090*	0.084*		0.005*	0.105*	0.115*
			(−0.129/−0.016)	(0.033/0.146)	(0.029/0.141)		(0.001/0.013)	(0.032/0.181)	(0.041/0.189)
Born before 1934 (ref.)									
1934–1943		0.435*			0.126*		−0.139*		−0.228*
		(0.398/0.567)			(0.079/0.190)		(−0.204/−0.090)		(−0.309/−0.156)
1944–1953		0.695*			0.201*		−0.222*		−0.364*
		(0.570/0.821)			(0.134/0.279)		(−0.301/−0.158)		(−0.450/−0.291)
1954–1963	0.088*	0.854*			0.275*		−0.212*		−0.487*
	(0.007/0.167)	(0.728/0.981)			(0.193/0.374)		(−0.317/−0.113)		(−0.583/−0.401)
1964–1973	0.269*	0.861*		−0.090*	0.246*	0.125*	−0.216*		−0.520*
	(0.189/0.348)	(0.743/0.983)		(−0.169/−0.018)	(0.149/0.360)	(0.034/0.221)	(−0.312/−0.120)		(−0.615/−0.434)
1974–1983	0.224*	0.896*		−0.181*	0.151*	0.178*	−0.164*		−0.528*
	(0.142/0.301)	(0.781/1.031)		(−0.265/−0.095)	(0.052/0.252)	(0.081/0.274)	(−0.271/−0.069)		(−0.625/−0.445)
Born after 1983	0.165*	0.793*		−0.373*		0.131*	−0.163*		−0.456*
	(0.073/0.261)	(0.651/0.928)		(−0.478/−0.276)		(0.020/0.242)	(−0.284/−0.054)		(−0.550/−0.368)
R^2	0.030	0.352	0.144	0.211	0.211	0.084	0.084	0.138	0.138
Model fit	CFI=.959; RMSEA=.038; SRMR=.029								

Notes: *p<0.05; unstandardized coefficients; bootstrapped confidence intervals below coefficient in parentheses; only significant coefficients are shown; negative coefficients on laws and activism indicate more acceptance of laws or more activism. N=2560.

Source: ISSP 2010, own calculations.

A direct effect on action is present in the three youngest cohorts; they act less environmentally friendly in their day-to-day activities and also support laws less strongly. A negative coefficient especially for the youngest group is also present in the other countries analyzed so far (i.e. the Czech Republic, Germany, Norway and New Zealand).

The total effects on all three types of action retrace the path of the effects of the cohorts on concern, although the large positive coefficient on evaluative concern and the negative direct effect on action for the cohort born after 1983 cancel each other out. Consequently, the total effect is not significant. The youngest generation in Spain was burdened with a relative high youth unemployment rate of 41.6% in 2010 (Eurostat 2011), which could explain why it is less engaged in environmental everyday action. However, the same tendency in the other countries indicates that it is not high unemployment alone but some other more universal causes that lead to reduced everyday action in this generation. This aspect will be discussed in the general summary.

General values (Table ES 5): In Spain, robust predictors of environmentalism are post-materialist worldviews for all three dimensions, especially in the cognitive and evaluative dimensions as well as for environmental activism. The signs of the coefficients appear in accordance with the hypothesis. As in the Czech Republic or New Zealand, Spanish materialists support laws more strongly than mixed types or post-materialists. This effect would mirror the argumentation applied in the Czech Republic (higher acceptance of laws in general, attention shifting; see chapter 7.2.2). When looking at the total effects, in all three types of action, mixed types and materialists are less engaged. The almost equally-sized coefficients of mixed types and materialists concerning the acceptance of laws result from the direct effect of the materialists that increases the support and the indirect effects via concern which decrease the support.

The political position's results are in harmony with the assumption that those on the left are stronger environmentalists. The far left perceive environmental problems somewhat more intensely than the rest. The far left and left react a little stronger to environmental issues on an evaluative level, non-voters are more willing to pay. Central voters are also more willing to pay but only on a 10% significance level; therefore the coefficient is omitted in Table ES 5. The effect of the non-voters, despite its significance, is quite small and lacks substantive practical relevance. The left voters support laws a little more strongly while the far-left voters are the second most engaged activists after the central voters. In total, those on the far lefts and the left are more active in all three aspects. The direct effect of the central voters on activism translates into a significant total effect.

Table ES 5: General values and environmentalism in Spain

	Cognitive	Evaluative	Conative	Action	Action (total)	Laws	Laws (total)	Activism	Activism (total)
Post-materialists (ref.)									
Mixed type	−0.130*	−0.286*			−0.107*		0.107*	0.198*	0.362*
	(−0.215/−0.039)	(−0.392/−0.197)			(−0.154/−0.070)		(0.071/0.151)	(0.077/0.320)	(0.242/0.477)
Materialists	−0.272*	−0.398*	−0.083*		−0.129*	−0.075*	0.103*	0.237*	0.498*
	(−0.365/−0.173)	(−0.515/−0.295)	(−0.147/−0.015)		(−0.208/−0.056)	(−0.146/−0.008)	(0.030/0.178)	(0.098/0.373)	(0.371/0.633)
No answer (ref.)									
Far left	0.188*	0.539*			0.184*		−0.193*	−0.374*	−0.656*
	(0.018/0.364)	(0.319/0.789)			(0.104/0.275)		(−0.288/−0.108)	(−0.609/−0.131)	(−0.912/−0.406)
Central/left		0.185*			0.043*	−0.087*	−0.141*		−0.077*
		(0.044/0.324)			(0.011/0.083)	(−0.168/−0.003)	(−0.229/−0.044)		(−0.145/−0.016)
Central								−0.517*	−0.540*
								(−0.801/−0.238)	(−0.826/−0.271)
Right/conservative									
Far right									
Non voter		0.151*	0.070*		0.041*		−0.050*		−0.070*
		(0.034/0.279)	(0.009/0.135)		(0.013/0.077)		(−0.091/−0..016)		(−0.128/−0.016)
Trust	−0.214*	0.422*	0.643*		0.080*		−0.112*	−0.575*	−0.767*
	(−0.382/−0.060)	(0.248/0.600)	(0.469/0.810)		(0.006/0.158)		(−0.189/−0.040)	(−0.791/−0.360)	(−0.971/−0.545)
R^2	0.024	0.261	0.172	0.163	0.163	0.079	0.079	0.157	0.157
Model fit	CFI=0.970; RMSEA=0.034; SRMR=0.027								

Notes: *p<0.05; unstandardized coefficients; bootstrapped confidence intervals below coefficient in parentheses; only significant coefficients are shown; negative coefficients on laws and activism indicate more acceptance of laws or more activism. N=2560.

Source: ISSP 2010, own calculations.

As in the other country samples, trust shows the same signs and relatively large effect sizes with a small but significant total effect on action and the support of laws. Environmental activism is moreover increased by trust directly which leads to a relatively large total effect.

Both the more rational perception of threats and the evaluative aspect of environmental concern are dependent on post-materialist values and to a lesser extent on political values. This supports the assumption taken from the constructivism theory that both value- and instrumental-rational drivers of action are prone to social construction via the general world views we obtain during the course of our life. To a lesser extent this is true for the support of laws and activism where some direct effects were encountered.

Social positions (Table ES 6): As for most other countries the frequency of people giving 'no answer' to the question about their education is very low (N=8) and therefore they become part of the reference group. The same procedure is pursued for the occupational position where only five persons refused to provide an answer.

With focus on the conative dimension, everyday action and the acceptance of laws social positions contribute little to their explanation via direct paths. All direct effects register a value of less than 0.1. Only education has a substantial direct impact on activism. Concerning the cognition and evaluation of environmental issues, social positions reveal a bit more. Even though richer people are less concerned regarding the cognitive dimensions, they exhibit more concern in the other two dimensions. In line with the wealth hypothesis, wealthier people are more concerned. The negative influence on the cognition hints at a lower exposition to environmental problems by wealthy people and thus lower scores on perception. As already discovered in Germany West, the income effect on the evaluative dimension is not linear. Taken together, at first more income increases concern in this dimension and, after a tipping point, it decreases it again.

Like in all other nations the effects of education are strongest. Higher secondary graduates, university incomplete and university graduates perceive environmental threats as slightly more dangerous. A higher degree leads to higher values in the evaluative dimension. Even people with lowest education are more concerned on the evaluative dimensions than people who do not hold any educational degree. This increase stops with those having an incomplete university degree which means that from the incomplete to the complete university degrees there is no further increase in the evaluative environmental concern. As the direct coefficients show, the three highest educational groups, in addition to the indirect effect via concern, are also more often engaged in activism.

Table ES 6: Social positions and environmentalism in Spain

	Cognitive	Evaluative	Conative	Action	Action (total)	Laws	Laws (total	Activism	Activism (total)
Equivalence income	−0.037*	0.079*	0.061*					−0.067*	−0.081*
	(−0.069/−0.007)	(0.037/0.126)	(0.028/0.090)					(−0.107/−0.025)	(−0.122/−0.039)
Equivalence income²		−0.018*							
		(−0.031/−0.006)							
No formal education (ref.)									
Lowest formal education		0.233*			0.063*		−0.076*		−0.105*
		(0.119/0.334)			(0.032/0.104)		(−0.117/−0.040)		(−0.159/−0.057)
Intermediate	0.125*	0.490*			0.170*		−0.182*		−0.262*
	(0.052/0.203)	(0.384/0.599)			(0.118/0.234)		(−0.239/−0.134)		(−0.338/−0.195)
Secondary	0.213*	0.588*		−0.071*	0.153*		−0.235*	−0.091*	−0.441*
	(0.132/0.298)	(0.478/0.707)		(−0.0137/−0.001)	(0.070/0.243)		(−0.298/−0.174)	(−0.118/0.001)	(−0.537/−0.330)
University incomplete	0.233*	0.738*			0.271*		−0.287*	−0.244*	−0.690*
	(0.120/0.349)	(0.607/0.881)			(0.197/0.369)		(−0.366/−0.211)	(−0.380/−.100)	(−0.826/−0.543)
University complete	0.195*	0.713*			0.254*		−0.275*	−0.452*	−0.923*
	(0.088/0.314)	(0.580/0.846)			(0.180/0.343)		(−0.351/−0.202)	(−0.578/−0.308)	(−1.044/−0.791)
No answer									
Never had paid work (ref.)									
No paid work	0.087*			0.064*	0.091*	−0.083*	−0.102*		−0.022*
	(0.011/0.172)			(−0.002/0.124)	(0.023/0.155)	(−0.153/−0.014)	(−0.175/−0.032)		(−0.047/−0.004)
In paid work	0.119*	0.078*			0.058*		−0.053*		−0.095*
	(0.043/0.198)	(0.017/0.136)			(0.029/0.088)		(−0.082/−0.028)		(−0.134/−0.060)
No answer									

	Cognitive	Evaluative	Conative	Action	Action (total)	Laws	Laws (total	Activism	Activism (total)
School children in household	0.038*				0.011*				
	(0.000/0.078)				(0.000/0.024)				
Toddlers in household									0.018*
									(0.008/0.029)
Big city (ref.)									
Suburbs		0.150*					−0.046*		−0.068*
		(0.045/0.258)					(−0.089/−0.015)		(−0.128/−0.025)
Town									0.008*
									(0.001/0.019)
Village	−0.053*	0.069*		−0.077*	−0.075*				
	(−0.011/0.004)	(0.11/0.127)		(−0.144/−0.013)	(−0.144/−0.012)				
Farm	−0.144*				−0.043*		0.032*		
	(−0.295/0.002)				(−0.094/0.001)		(−0.001/0.069)		
No answer									
R^2	0.035	0.389	0.149	0.174	0.174	0.080	0.080	0.171	0.171
Model fit	CFI=0.967; RMSEA=0.030; SRMR=0.025								

Notes: *$p<0.05$; unstandardized coefficients; bootstrapped confidence intervals below coefficient in parentheses; only significant coefficients are shown; negative coefficients on laws and activism indicate more acceptance of laws or more activism. N=2560.

Source: ISSP 2010, own calculations.

Table ES 7: Full model Spain

	Cognitive	Evaluative	Conative	Action	Action (total)	Laws	Laws (total)	Activism	Activism (total)
Gender		0.050*	−0.070*	0.083*	0.093*			0.111*	0.095*
		(0.002/0.115)	(−0.132/−0.007)	(0.025/0.148)	(0.037/0.154)			(0.043/0.186)	(0.025/0.165)
Born before 1934 (ref.)									
1934–1943		0.368*			0.109*		−0.134*		−0.159*
		(0.243/0.497)			(0.060/0.170)		(−0.201/−0.079)		(−0.228/−0.098)
1944–1953		0.496*			0.147*		−0.180*	0.117*	−0.099*
		(0.368/0.631)			(0.086/0.217)		(−0.255/−0.124)	(0.015/0.209)	(−0.204/−0.003)
1954–1963		0.574*			0.169*		−0.124*		−0.253*
		(0.430/0.704)			(0.108/0.244)		(−0.223/−0.026)		(−0.335/−0.179)
1964–1973	0.174*	0.549*		−0.108*	0.109*	0.125*	−0.118*	0.118*	−0.166*
	(0.100/0.252)	(0.413/0.685)		(−0.193/−0.027)	(0.016/0.218)	(0.030/0.230)	(−0.219/−0.015)	(0.009/0.227)	(−0.284/−0.056)
1974–1983	0.117*	0.575*		−0.203*		0.181*		0.152*	−0.129*
	(0.036/0.196)	(0.436/0.711)		(−0.285/−0.116)		(0.076/0.280)		(0.027/0.272)	(−0.248/−0.005)
Born after 1983		0.508*		−0.397*	−0.222*	0.129*		0.129*	
		(0.356/0.651)		(−0.500/−0.296)	(−0.325/−0.120)	(0.032/0.246)		(0.003/0.246)	
Post-materialists (ref.)									
Middle	−0.109*	−0.191*			−0.092*	−0.094*		0.142*	0.258*
	(0.,211/−0.017)	(−0.277/−0.102)			(−0.143/−0.053)	(−0.188/0.000)		(0.001/0.289)	(0.109/0.412)
Materialists	−0.213*	−0.198*			−0.132*	−0.147*		0.162*	0.314*
	(−0.320/−0.118)	(−0.294/−0.102)			(−0.185/−0.085)	(−0.259/−0.038)		(0.001/0.315)	(0.148/0.466)
No answer (ref.)									
Far left	0.178*	0.404*			0.177*		−0.195*	−0.334*	−0.554*
	(−0.014/0.358)	(0.181/0.625)			(0.087/0.279)		(−0.298/−0.100)	(−0.633/−0.028)	(−0.879/−0.235)
Left		0.184*			0.075*	−0.104*	−0.191*		−0.096*
		(0.081/0.303)			(0.039/0.121)	(−0.205/−0.104)	(−0.308/−0.100)		(−0.157/−0.042)
Central								−0.416*	−0.416*

Table ES 7: Continued

	Cognitive	Evaluative	Conative	Action	Action (total)	Laws	Laws (total)	Activism	Activism (total)
								(−0.826/−0.074)	(−0.826/−0.074)
Right/Conservative		0.127*			0.038*		−0.053*		−0.050*
		(0.016/0.250)			(0.004/0.080)		(−0.107/−0.010)		(−0.110/0.002)
Far right									
Non-voter		0.098*	0.070*		0.034*		−0.096*		
		(0.003/0.204)	(0.015/0.136)		(0.007/0.069)		(−0.191/−0.019)		
Trust	−0.271*	0.246*	0.590*					−0.411*	−0.503*
	(−0.431/−0.110)	(0.079/0.410)	(0.412/0.768)					(−0.610/−0.210)	(−0.708/−0.310)
Equivalence income	−0.033*	0.036*	0.043*						
	(−0.068/0.000)	(0.001/0.068)	(0.014/0.072)						
Equivalence income ^{2}		−0.013*			−0.004*				
		(−0.025/0.002)			(−0.008/0.000)				
No formal education (ref.)									
Lowest formal education		0.113*					−0.044*		−0.050*
		(0.004/0.225)					(−0.091/−0.005)		(−0.100/−0.002)
Intermediate		0.266*		0.083*	0.163*		−0.104*		−0.127*
		(0.152/0.397)		(0.016/0.153)	(0.089/0.257)		(−0.166/−0.059)		(−0.198/−0.068)
Higher secondary	0.103*	0.345*			0.137*		−0.163*	−0.119*	−0.310*
	(0.027/0.178)	(0.214/0.472)			(0.082/0.207)		(−0.233/−0.105)	(−0.222/−0.017)	(−0.419/−0.204)
University incomplete	0.131*	0.474*		0.105*	0.289*		−0.221*	−0.294*	−0.555*
	(0.024/0.237)	(0.328/0.619)		(0.005/0.203)	(0.179/0.400)		(−0.301/−0.152)	(−0.466/−0.133)	(−0.737/−0.385)
University	0.098*	0.441*			0.166*		−0.202*	−0.493*	−0.737*
	(0.005/0.204)	(0.304/0.582)			(0.102/0.249)		(−0.279/−0.136)	(−0.661/−0.329)	(−0.909/−0.572)
Never had paid work (ref.)									

Table ES 7: Continued

	Cognitive	Evaluative	Conative	Action	Action (total)	Laws	Laws (total)	Activism	Activism (total)
No paid work					0.024*		−0.018*		
					(0.000/0.053)		(−0.041/0.000)		
In paid work	0.084*				0.028*		−0.025*		−0.028*
	(0.002/0.156)				(0.002/0.052)		(−0.043/−0.006)		(−0.046/−0.008)
School children in household									
Toddlers in household									
Big city (ref.)									
Suburbs		0.127*			0.038*		−0.047*		−0.060*
		(0.027/0.234)			(0.009/0.075)		(−0.090/−0.010)		(−0.124/−0.018)
Town				−0.081*	−0.081*				0.002*
				(−0.162/0.002)	(−0.162/0.002)				(0.000/0.006)
Village				−0.085*	−0.070*				
				(−0.158/−0.018)	(−0.145/−0.006)				
Farm									
NA community size									
Cognitive		0.347*		0.216*	0.319*	−0.104*	−0.233*	−0.076*	−0.234*
		(0.281/0.417)		(0.150/0.281)	(0.258/0.390)	(−0.170/−0.031)	(−0.291/−0.170)	(−0.153/−0.007)	(−0.309/−0.171)
Evaluative			0.433*	0.266*	0.296*	−0.344*	−0.372*	−0.417*	−0.453*
			(0.360/0.526)	(0.166/0.381)	(0.199/0.409)	(−0.458/−0.250)	(−0.484/−0.283)	(−0.546/−0.289)	(−0.577/−0.336)
Conative				0.070*	0.070*	−0.064*	−0.064*	−0.083*	−0.083*
				(0.021/0.118)	(0.021/0.118)	(−0.119/−0.006)	(−0.119/−0.006)	(−0.136/−0.026)	(−0.136/−0.026)
R^2	0.059	0.459	0.183	0.222	0.222	0.093	0.093	0.191	0.191
Model fit	CFI=0.977; RMSEA=0.025; SRMR=0.0								

Notes: *p<0.05; unstandardized coefficients; bootstrapped confidence intervals below coefficient in parentheses; only significant coefficients are shown; negative coefficients on laws and activism indicate more acceptance of laws or more activism. N=2560.

Source: ISSP 2010, own calculations.

The other positional factors, again as in most other countries, are of minor importance, despite some small but significant coefficients. People in employment and individuals temporarily without a job are a bit more concerned (cognitive), and the first group also exhibits higher values in the evaluative dimension. Inhabitants of the suburbs and villages are marginally more concerned (evaluative) while those living in a village or on a farm perceive environmental problems as less dangerous.

As there are barely any direct effects, and, except for the effects of education on activism, the coefficients are very small, the total effects resemble the impact on environmental attitudes. Yet, only the effects of education are of some practical relevance, the others are simply too small.

Full model: Besides the usual changes in coefficient sizes (and some losses of significance), some important differences are found in Table ES 7. While virtually all coefficients of the birth cohorts dropped in size, in the full model direct effects on activism appear due to the inclusion of general values. With the loss of significance of the 1984+ cohort on the cognitive dimension, the total negative effect becomes significant.

The increase in size of the materialists' and mixed types' direct impact on the acceptance of laws negates the total effects which are present in Table ES 5. Another important effect appears among the general values: right/conservatives show an increased evaluative concern which leads to minimally higher values in everyday action, laws and activism. However, they differ less from the reference category than from the far-left and left. This effect underlines that conservative world views seem to be quite versatile and when controlling for birth cohorts and education, only a positive effect remains.

When controlled for the birth cohorts, quite large drops in the educational effects on evaluative concern appear; they nevertheless remain significant and of considerable size.

The coefficients of the occupational position are mostly accounted for when the birth cohorts are included for the same reasons as in other nations, i.e. those without a job at the moment tend to be old and retired, those with a job are of middle age, and individuals that never had a job so far are mostly young people. A small effect of having a job on the cognition of threats remains and most total effects (in part mediated via income), though very small, also persist.

As in most cases, among the three components of environmental concern the evaluative dimension is best explained by the introduced variables, in particular by the birth cohorts and social positions. Nearly 50% of the variation in the evaluation of environmental issues is due to social construction. The perception of environmental threats, in contrast, is much less affected by the variables in the model. With an R^2 of only 0.059 the perception of threats is a relatively universal issue. Because the evaluative dimension is also the most consistent predictor of all types of action in Spain, indirect increases in environmental action can be gained when addressing the cohorts' differences and improving education. Thus, in particular everyday action and activism can be increased. With regard to the acceptance of laws, other aspects need to be found that are able to account for the variance in Spain.

From continental Europe the journey now continues to another part of the world. On the North American subcontinent the environmentalism of the United States of America now is the subject of our attention as the last country of analysis.

6.6 USA

GPD per capita (2010): $46,612

People in the United States of America are said to have the reputation of being less environmentally concerned than those in other countries as discussed by Dietz et al. (2007) with regard to support for climate change policy. Hadler and Haller (2011), for instance, analyze the ISSP 2000 and report descriptive statistics for private (or everyday) environmental action.[21] Comparing the nations that are analyzed in the present study and also investigated by the authors, the USA indeed reveals the lowest amount of action.[22] Including all countries available in the ISSP 2000 the USA ranks 14 out of 23 in private behaviors and 8th in activism. In Diekmann and Franzen (1999) the USA can be found in an upper middle concerned position in an index of rating items. Ranking the environment in relation to other issues, the USA occupies a middle position. The upper middle position is retained in Franzen and Meyer (2010) in a multi-item index of environmental concern. The same is true for the USA's willingness to pay: the country holds an upper middle position in Nawrotzki (2012). The USA ranks 26 out of 39 countries concerning their percentage of support for environmental protection in an earlier study by Brechin (1999) using the World Values Survey. In the Inglehart study (1995) the USA's willingness to pay occupies position 21 out of 39. At the end of the day, people in the USA seem

---

21 In the 2000 module only the items on the frequency of recycling and cutting back car driving were available.

22 The nations are: Austria, Canada, the Czech Republic, Finland, Germany West and East, Great Britain, New Zealand, Norway, Spain, Sweden, and the USA.

to be averagely environmentally concerned. In the present study the picture is varied. Although the perception of threats and evaluative concern only have the twelfth-highest value (Table 5.1), the USA's willingness to pay is third highest. However, everyday action is 5th lowest, activism is 7th lowest and the acceptance of laws is actually lowest in the USA. Taken altogether in the ISSP 2010 the USA ranks as a country with less pronounced environmentalism.

As is the case for the Czech Republic, Germany East, New Zealand and Spain the economy is most important to US respondents while the environment is ranked fifth (Table USA 1). This is consistent with the average of the six countries. A major difference to the other nations is the position of terrorism. In the USA it is judged to be the fourth-most important problem whereas across all countries it only ranks eighth.

Table USA 1: Most important issues in the USA

	Most important	Second most important	Score
The economy	492	203	593,5
Health care	309	327	472,5
Education	258	325	420,5
Terrorism	107	152	183
The environment	58	120	118
Immigration	53	76	91
Poverty	37	71	72,5
Crime	19	73	55,5
None of these	13	6	16

Notes: Source: ISSP 2010, own calculations.

Even though climate change is the second-most important environmental problem across the six countries, in the more climate-skeptical USA, it only ranks fifth (Table USA 2). This is in line with lower support for climate policy in the USA than in other countries as discussed above (Dietz *et al.* 2007). The other issues that are of a certain universal importance, i.e. air pollution, using up our natural resources and water pollution are also among the most important ones in the USA. When checking the six nations' individual scores against each other, many differences in their priorities are obvious. In comparison to Germany, where nuclear waste occupies second and third position in West and East respectively, people in the USA consider it least serious. While the air seems to be of quite good quality in New Zealand (third last), the USA considers air pollution the most important problem in particular for the family. Thus, like in every country, the USA exhibits their individual perception of environmental problems.

Table USA 2: Most important environmental issue in the USA

	For country	For family	Score
Air pollution	199	300	250
Using up our natural resources	344	150	247
Chemicals and pesticides	112	159	136
Water pollution	134	124	129
Climate change	127	85	106
Genetically modified foods	61	132	97
Water shortage	95	71	83
None of these	22	123	73
Domestic waste disposal	60	65	63
Nuclear waste	88	15	52

Notes: Source: ISSP 2010, own calculations.

Regarding their future energy supply (Table USA 3) the USA also deviates from the average. Like in all countries, hardly surprisingly solar, water and wind energy are preferred, although less strongly in the USA where only 53.8% consider the renewables the best option. On average, as well as in the Czech Republic, New Zealand, Spain and Norway, the second most important source is nuclear energy. As in Germany West, it is least important in the USA (leaving aside the 'none of these' category). While on average fuels made from crops are preferred over fossil fuels, the USA vote is the other way round. Peoples' preference is in stark contrast to the actual situation. Coal, oil and natural gas combined with nuclear power constitute 94.0% of the USA's primary energy supply and solar, wind and water power only make up 1.8%.

A brief investigation of tables USA 1–3, however, does not indicate that people in the USA are substantially less environmentally concerned than in the other five countries.

As can be read from tables 5.2–5.4 attitudes have a relatively strong impact on the three types of action. This also means that the effects of the cohorts, gender, general values and social positions are translated to a greater degree into action. Starting with cohorts and gender, the multivariate analysis will unveil these effects.

Table USA 3: Energy supply in the USA

	Best source for future energy supply		Real total primary energy supply (TPES) 2010
	%	N	%
Solar, wind or water power	53.8%	650	1.8%
Fuels made from crops	18.1%	219	4.1%
Coal, oil and natural gas	14.1%	170	84.1%
Nuclear power	11.9%	144	9.9%
None of these	2.1%	25	0.1%
Total	100%	1208	100%

Notes: Source: ISSP 2010, own calculations. International Energy Agency (2012).

Multivariate analysis

Base model: CFI=0.951; RMSEA=0.057; SRMR=0.045; Cognitive → affective=0.421 (0.360/0.492); cognitive → conative=0.167 (0.105/0.228); affective → conative=0.465 (0.386/0.557).

Cohorts and gender (Table USA 4): The results for gender are quite similar to most other country samples. Women are more concerned and act marginally more benignly in their day-to-day routine. At the same time they support laws less strongly than men and are less often engaged in activism. The total effects are only significant concerning action and activism. The higher environmental concern of women and their lower support for laws cancel each other out.

The four birth cohorts born from 1934 to 1973 are mostly concerned (cognitive) to a relatively constant degree. A slightly lower coefficient is found for the cohorts 1974–83 and 1984+. The two generations born between 1964 and 1983 exhibit the highest degree of evaluative concern. With regard to action, the older cohorts are more active.

Table USA 4: Cohorts, gender and environmentalism in the USA

	Cognitive	Evaluative	Conative	Action	Action (total)	Laws	Laws (total)	Activism	Activism (total)
Male (ref.)									
Female	0.174*			0.074*	0.108*	0.132*		0.149*	0.108*
	(0.090/0.262)			(0.004/0.139)	(0.045/0.165)	(0.025/0.232)		(0.048/0.239)	(0.007/0.204)
Born before 1934 (ref.)									
1934–1943	0.326*			0.156*	0.135*	−0.226*	−0.102*		−0.076*
	(0.108/0.536)			(0.046/0.272)	(0.064/0.199)	(−0.424/−0.016)	(−0.160/−0.037)		(−0.133/−0.029)
1944–1953	0.315*			0.101*	0.124*	−0.185*	−0.106*	−0.171*	−0.244*
	(0.128/0.516)			(0.107/0.193)	(0.056/0.192)	(−0.363/−0.028)	(−0.169/−0.047)	(−0.298/−0.043)	(−0.375/−0.102)
1954–1963	0.379*				0.073*	−0.267*	−0149*		−0.089*
	(0.196/0.571)				(0.037/0.114)	(−0.426/−0.101)	(−0.213/−0.086)		(−0.144/−0.043)
1964–1973	0.328*	0.108*			0.074*		−0.135*		−0.129*
	(0.146/0.522)	(0.005/0.217)			(0.037/0.115)		(−0.200/−0.064)		(−0.207/−0.064)
1974–1983	0.275*	0.098*			0.060*		−0.112*		−0.113*
	(0.077/0.459)	(−0.003/0.208)			(0.022/0.099)		(−0.176/−0.045)		(−0.188/−0.044)
Born after 1983	0.276*				0.047*		−0.036*		−0.065*
	(0.079/0.469)				(0.014/0.085)		(−0.065/−0.010)		(−0.116/−0.017)
R^2	0.024	0.229	0.321	0.258	0.258	0.240	0.240	0.150	0.150
Model fit	CFI=.958; RMSEA=.041; SRMR=.034								

Notes: *p<0.05; unstandardized coefficients; bootstrapped confidence intervals below coefficient in parentheses; only significant coefficients are shown; negative coefficients on laws and activism indicate more acceptance of laws or more activism. N= 1430.

Source: ISSP 2010, own calculations.

The 1934–53 cohorts show more everyday action and support for laws, which is also higher among people born between 1954 and 63. A considerably more frequent involvement in activism is only present in the 1944–53 generation. The combination of the effects on concern and the direct effects on the three activities leads to total effects which are not overly large in size.

General values (Table USA 5): Postmaterialist values do not play a role in the USA in the formation of environmental attitudes, as the empty cells in Table USA 5 show. As encountered before (e.g. in New Zealand), mixed types and materialists are less likely activists but materialists support laws more strongly. An explanation for the higher acceptance of laws was provided in the profile of the Czech Republic (higher acceptance of laws in general, attention shifting) and is discussed in chapter 7.2.2.

The pattern of trust is the same as for the other countries. No direct effects on the three activities are found but all total effects are significant and increase the respective conduct with the strongest impact on the acceptance of laws.

Of little surprise but of most relevance is the more intense perception of environmental threats by strong democrats, democrats, independent-close to democrats as well as independent voters; especially in comparison to republican voters, whose coefficients decline the closer they move to the right. However, only the strong republican's coefficient is significant. Additionally, democrats and independent democrats exhibit a higher level of concern when evaluating environmental issues while strong republicans are less concerned. The higher environmental concern of the democrats, independent democrats and the independent voters is balanced by their lower everyday action. Correspondingly, their total effects on everyday action are not significant and only strong democrats are more active while the republicans and strong republicans are less so. The support for laws is straightforward at first sight. Strong democrats, democrats and independent-close democrats support laws more than non-voters or independent and independent-close republicans. The difference to non-voters is not surprising. Non-voters have a reason why they do not vote: less confidence in politics leads to a generally lower acceptance of legal measures than those who support the two established parties. The same can be argued for the independent voters. Astonishingly, republicans and also strong republicans prefer stricter laws over voluntary environmental protection than non-voters. The effect of the strong republicans is negated by their lower concern but the total effect of the republicans remains significant. On activism the political position exercises only a minor direct impact: independent and republican voters are less active.

Table USA 5: General values and environmentalism in the USA

	Cognitive	Evaluative	Conative	Action	Action (total)	Laws	Laws (total)	Activism	Activism (total)
Post-materialists (ref.)									
Middle position								0.146*	0.146*
								(0.006/0.286)	(0.006/0.286)
Materialists						−0.205*	−0.205*	0.290*	0.290*
						(−0.350/−0.067)	(−0.350/−0.067)	(0.115/0.458)	(0.115/0.458)
Other party/no answer (ref.)[a]									
Strong democrat	0.605*				0.230*	−0.215*	−0.464*		−0.134*
	(0.465/0.731)				(0.168/0.297)	(−0.380/−0.048)	(−0.638/−0.287)		(−0.177/−0.100)
Democrat	0.425*	0.132*		−0.165*		−0.196*	−0.455*		
	(0.301/0.553)	(0.029/0.235)		(−0.300/−0.053)		(−0.383/−0.031)	(−0.651/−0.279)		
Independent close democrat	0.482*	0.238*		−0.138*		−0.209*	−0.582*		−0.231*
	(0.340/0.631)	(0.121/0.362)		(−0.281/0.000)		(−0.398/−0.027)	(−0.778/−0.383)		(−0.306/−0.159)
Independent	0.210*			−0.122*			−0.086*	0.149*	
	(0.083/0.343)			(−0.242/0.002)			(−0.147/−0.037)	(0.021/0.287)	
Independent close republican									
Republican				−0.167*	−0.167*	−0.203*	−0.203*	0.166*	0.166*
				(−0.301/−0.034)	(−0.301/−0.034)	(−0.388/−0.028)	(−0.388/−0.028)	(0.029/0.304)	(0.029/0.304)
Strong republican	−0.218*	−0.206*			−0.114	−0.299*			0.143*
	(−0.379/−0.067)	(−0.346/−0.073)			(−0.180/−0.050)	(−0.515/−0.100)			(0.070/0.225)
Trust	−0.305*	1.097*	0.251*		0.136*		−0.631*		−0.510*
	(−0.516/−0.094)	(0.890/1.315)	(0.049/0.455)		(0.021/0.258)		(−0.820/−0.461)		(−0.645/−0.395)
R^2	0.125	0.356	0.331	0.255	0.255	0.242	0.242	0.149	0.149
Model fit	CFI=0.962; RMSEA=0.041; SRMR=0.035								

Notes: *$p<0.05$; unstandardized coefficients; bootstrapped confidence intervals below coefficient in parentheses; only significant coefficients are shown; negative coefficients on laws and activism indicate more acceptance of laws or more activism. N= 1430.

Source: ISSP 2010, own calculations.

Table USA 6: Social positions and environmentalism in the USA

	Cognitive	Evaluative	Conative	Action	Action (total)	Laws	Laws (total)	Activism	Activism (total)
Equivalence income	−0.160*	0.069*			−0.030*			−0.085*	−0.087*
	(−0.220/−0.104)	(0.024/0.116)			(−0.048/−0.010)			(−0.137/−0.031)	(−0.135/−0.028)
Equivalence income^2	0.042*								
	(0.018/0.076)								
No formal education (ref.)									
Lowest education						−0.268*	−0.268*		
						(−0.505/0.001)	(−0.505/0.001)		
Intermediate									
Secondary	−0.097*	0.332*	−0.171*				−0.120*		−0.105*
	(−0.191/0.004)	(0.202/0.461)	(−0.249/−0.089)				(−0.210/−0.031)		(−0.175/−0.044)
University incomplete		0.452*			0.056*		−0.381*		−0.195*
		(0.312/0.577)			(0.932/0.083)		(−0.526/−0.236)		(−0.274/−0.132)
University complete		0.639*		0.178*	0.238*	−0.237*	−0.601*	−0.257*	−0.585*
		(0.495/0.788)		(0.080/0.278)	(0.137/0.338)	(−0.371/−0.093)	(−0.748/−0.459)	(−0.375/−0.134)	(−0.704/−0.473)
Schoolchildren in household		−0.029*			−0.004*		0.016*		0.012*
		(−0.055/−0.008)			(−0.008/−0.001)		(0.003/0.032)		(0.003/0.023)
Toddlers in household								0.088*	0.099*
								(0.020/0.156)	(0.030/0.167)
Never had paid work (ref.)									
No paid work						−0.244*	−0.244*		
						(−0.479/0.00)	(−0.479/0.00)		
In paid work		0.089*				−0.241*	−0.291*		−0.051*
		(0.008/0.179)				(−0.466/−0.001)	(−0.512/−0.053)		(−0.087/−0.017)
No answer									
Big city (ref.)									

Table USA 6: Continued

	Cognitive	Evaluative	Conative	Action	Action (total)	Laws	Laws (total)	Activism	Activism (total)
Suburbs									
Town	−0.127*				−0.043*		0.056*	0.141*	0.185*
	(−0.245/−0.014)				(−0.097/0.000)		(0.006/0.113)	(0.015/0.266)	(0.059/0.313)
Village	n.a.	n.a.	n.a.	n.a.	n.a.	n.a.	n.a.	n.a.	n.a.
Farm	n.a.	n.a.	n.a.	n.a.	n.a.	n.a.	n.a.	n.a.	n.a.
R^2	0.031	0.357	0.328	0.259	0.259	0.244	0.244	0.180	0.180
Model fit	CFI=.972; RMSEA=.035; SRMR=.033								

Notes: *p<0.05; unstandardized coefficients; bootstrapped confidence intervals below coefficient in parentheses; only significant coefficients are shown; negative coefficients on laws and activism indicate more acceptance of laws or more activism. N= 1430.

Source: ISSP 2010, own calculations.

In total, only strong democrats and independent democrats are more involved in activism, at the same time republicans are less involved with regard to the total effects.

Social positions (Table USA 6): For the USA, the categories 'no answer' for education, occupational situation and community size are included in the reference category due to frequencies lower or equal to five. A special feature of the USA's sample is that obviously only the urban population was surveyed as not a single person appears in the 'village' or 'farm' category. Consequently, the sample does not represent the whole population of the USA but only the urban part of the country.

The way income impacts environmental concern has been encountered before (e.g. Spain): wealthier people perceive environmental problems to be less dramatic. However, their evaluative reaction is stronger than the reaction of less wealthy people, resulting in a small significant total effect on action, and a non-significant one on laws. In line with expectation, wealthier people are more engaged in activism, but not very much more. The negative influence on the cognitive dimension is not linear as the significant coefficient of the squared equivalence income shows. Although a higher income leads to a lower score in the perception of threats, the curve follows a convex path which means that, from a certain point, the score increases again.

Particularly the evaluative dimension of environmental concern is positively influenced by education. However, the path from the evaluative dimension to everyday action is the weakest among the three dimensions but the strongest on laws and activism. As a result, the total effects of the three highest educational groups on laws and activism and those of the two highest groups on everyday action are significant. The total effect of university graduates is boosted by a direct impact on the three types of action. Conversely, the stronger predictors of everyday action (cognition of environmental threats) are negatively affected by holding a higher secondary degree. This negates their total effect on action and lowers their effect on laws and activism. Running counter to assumption, individuals with the lowest formal education accept laws more strongly. One possible explanation would be that at this point the low education entails a lack of knowledge and capabilities to decide how to protect the environment; hence, it is the government's responsibility to protect the environment. This reasoning is not fully satisfactory because one would assume the same to be true for those without any education but this group supports laws less strongly.

The only significant coefficient regarding the occupational position on environmental attitudes is the effect of employed people on the evaluative dimension, in part mediated via income. It does not significantly affect everyday action, but having a job increases activism. Although these effects are very small, the acceptance of laws is directly influenced by having a job and currently having no job to a more substantial degree. The stable position given by being employed leads people to evaluate environmental issues in a more environmentally friendly way, and be more involved in activism. As in other country samples, those having no paid work at the moment are mostly older people and a decline of this coefficient is expected in the full model. In a similar way to the other countries, the variables derived from the biographical availability assumption do not contribute remarkably to the explanation of environmental attitudes, everyday action or activism in the USA. Only the increase of the acceptance of laws may be in line with the argument that the lack of time that goes with having a job makes people prefer the government to arrange environmental protection.

Finally, people living in a town perceive threats less strongly, maybe due to less exposure to them in comparison to big city and suburban citizens. This leads to less action and acceptance of laws. Lower concern has a direct effect on lowering activism in towns. It could be argued that many forms of activism (e.g. a demonstration) are less available in smaller communities.

Final model (Table USA 7): Comparing Table USA 7 to the tables of the separate analyses (USA 4–6), many notable changes occur. Unlike in the other nations, where the cohort effects mostly remained stable, in the USA many lost significance with the introduction of the general value measures. Especially striking is the loss of significance of the total effects on activism. Much of the generational differences are instead hidden in ideological disparities.

For the reason just mentioned, various differences are also found in the political position. Summed up across all coefficients, the political position yields a clearer picture than in the separate analysis of general values. In the full model, strong republicans, republicans, independent-but-close republicans all show a negative significant coefficient on the evaluative dimension while at least the positive effect for the independent democrats remains significant. The negative effects on action of independent democrats and independents disappear as do the effects on laws for the same two groups plus democrats. This leads to a more pronounced and thus to a clearer picture with more action in all respects on the democratic side and less action on the republican side.

Table USA 7: Full model USA

	Cognitive	Evaluative	Conative	Action	Action (total)	Laws	Laws (total)	Activism	Activism (total)
Gender	0.110*			0.093*	0.139*		–0.043*	0.114*	
	(0.027/0.194)			(0.006/0.179)	(0.046/0.225)		(–0.082/–0.007)	(0.009/0.204)	
Born before 1934 (ref.)									
1934–1943	0.241*			0.151*	0.249*		–0.100*		
	(0.055/0.426)			(0.013/0.297)	(0.101/0.403)		(–0.182/–0.018)		
1944–1953	0.270*	–0.176*			0.085*				
	(0.098/0.441)	(–0.281/–0.073)			(0.016/0.161)				
1954–1963	0.302*						–0.126*		
	(0.128/0.464)						(–0.205/–0.051)		
1964–1973	0.258*				0.107*		–0.106*		
	(0.082/0.419)				(0.039/0.181)		(–0.186/–0.030)		
1974–1983	0.179*				0.080*				
	(0.004/0.353)				(0.016/0.152)				
Born after 1983									
Post-materialists (ref.)									
Middle									
Materialists						–0.238*	–0.238*	0.247*	0.247*
						(–0.370/–0.113)	(–0.370/–0.113)	(0.099/0.409)	(0.099/0.409)
No answer (ref.)									
Strong democrat	0.554*				0.218*		–0.232*		–0.119*
	(0.425/0.691)				(0.159/0.282)		(–0.306/–0.168)		(–0.186/–0.067)
Democrat	0.393*			–0.125*			–0.164*		–0.072*

Table USA 7: Continued

	Cognitive	Evaluative	Conative	Action	Action (total)	Laws	Laws (total)	Activism	Activism (total)
	(0.268/0.534)			(–0.223/–0.004)			(–0.239/–0.106)		(–0.135/–0.030)
Independent democrat	0.452*	0.167*			0.135*		–0.304*		–0.163*
	(0.320/0.591)	(0.053/0.284)			(0.000/0.262)		(–0.348/–0.211)		(–0.232/–0.100)
Independent	0.175*				0.072*		–0.072*		
	(0.043/0.306)				(0.023/0.126)		(–0.127/–0.017)		
Independent republican		–0.190*			–0.022*		0.104*		0.091*
		(–0.313/–0.043)			(–0.044/–0.003)		(0.026/0.182)		(0.040/0.157)
Republican		–0.173*		–0.128*	–0.147*		0.092*	0.120*	0.203*
		(–0.281/–0.054)		(–0.260/0.000)	(–0.279/–0.013)		(0.027/0.156)	(0.000/0.231)	(0.079/0.315)
Strong republican	–0.212*	–0.364*			–0.128*		0.281*		0.193*
	(–0.375/–0.045)	(–0.485/–0.232)			(–0.194/–0.056)		(0.171/0.388)		(0.119/0.217)
Trust	–0.274*	0.868*	00.226*				–0.405*		–0.317*
	(–0.485/–0.011)	(0.675/10.067)	(–00.001/00.0451)				(–569/–0.244)		(–0.466/–0.184)
Equivalence income	–0.114*	0.059*			–0.016*			–0.080*	–0.087*
	(–0.175/–0.055)	(0.014/0.102)			(–0.036/0.002)			(–0.138/–0.019)	(–0.145/–0.026)
Equivalence income²	0.038*				0.015*				
	(0.011/0.067)				(0.004/0.026)				
No formal education (ref.)									
Lowest formal education									0.030*

Table USA 7: Continued

	Cognitive	Evaluative	Conative	Action	Action (total)	Laws	Laws (total)	Activism	Activism (total)
									(0.009/0.058)
Intermediate									0.026*
									(0.008/0.053)
Higher secondary	−0.097*	0.361*	−0.160*				−0.122*		−0.113*
	(−0.181/−0.001)	(0.240/0.487)	(−0.241/−0.087)				(−0.217/−0.044)		(−0.180/−0.058)
University incomplete		0.464*					−0.362*		−0.180*
		(0.332/0.595)					(−0.516/−0.212)		(−0.254/−0.115)
University		0.606*		0.398*	0.464*	−0.237*	−0.568*	−0.258*	−0.545*
		(0.471/0.749)		(0.056/0.742)	(0.123/0.822)	(−0.383/−0.095)	(−0.710/−0.431)	(−0.389/−0.120)	(−0.677/−0.410)
Never had paid work (ref.)									
No paid work						−0.259*	−0.259*		
						(−0.520/0.000)	(−0.520/0.000)		
In paid work						−0.252*	−0.252*		−0.050*
						(−0.507/0.000)	(−0.507/0.000)		(−0.090/−0.019)
NA work									
School children in household									0.013*
									(0.004/0.024)
Toddlers in household								0.090*	0.090*
								(0.042/0.131)	(0.042/0.131)
Big city (ref.)									
Suburbs									

Table USA 7: Continued

	Cognitive	Evaluative	Conative	Action	Action (total)	Laws	Laws (total)	Activism	Activism (total)
Town								0.138*	0.156*
								(0.034/0.229)	(0.055/0.245)
Village	n.a.	n.a.	n.a.	n.a.	n.a.	n.a.	n.a.	n.a.	n.a.
Farm	n.a.	n.a.	n.a.	n.a.	n.a.	n.a.	n.a.	n.a.	n.a.
Cognitive		0.418*	0.189*	0.270*	0.385*	−0.157*	−0.420*	−0.094*	−0.258*
		(0.350/0.483)	(0.109/0.260)	(0.197/0.347)	(0.317/0.468)	(−0.244/−0.063)	(−0.498/−0.341)	(−0.177/−0.010)	(−0.332/−0.181)
Evaluative			0.399*		0.128*	−0.472*	−0.543*	−0.392*	−0.392*
			(0.311/0.503)		(0.090/0.178)	(−0.599/−0.335)	(−0.654/−0.423)	(−0.513/−0.286)	(−0.513/−0.286)
Conative				0.322*	0.322*	−0.184*	−0.184*		
				(0.240/0.405)	(0.240/0.405)	(−0.283/−0.076)	(−0.283/−0.076)		
R^2	0.150	0.466	0.335	0.283	0.283	0.244	0.244	0.188	0.188
Model fit		CFI=0.978; RMSEA=0.028; SRMR=0.025							

Notes: *p<0.05; unstandardized coefficients; bootstrapped confidence intervals below coefficient in parentheses; only significant coefficients are shown; negative coefficients on laws and activism indicate more acceptance of laws or more activism. N= 1430.

Source: ISSP 2010, own calculations.

Trust's total effect on action disappears because the positive effects on concern (evaluative and conative) drop more than the negative effect on the cognitive dimension. The total effects on the other actions remain significant.

The significant effects of the equivalence income and the squared equivalence income on the cognitive dimension remain in the full model.

For education it is notable that the size of the coefficient of a university degree approximately doubles in the full model. Most of this increase is traced back to the political positions and trust (which is higher among university graduates).

For the occupational position the small coefficient on evaluative concern disappears but the direct and total impact on laws remains nearly the same, even under control of the age composition.

For children and toddlers in the household only the coefficients that reduce people's activism remain. The same is true for people from towns that are still less likely to be activists but no longer less concerned. An investigation of the correlational patterns among town and the other predictors does not unveil specific reasons.

The findings from the USA are most interesting from a constructivist perspective because the political position especially plays a role for the mere perception of environmental issues. 12.5% of its variance is attributed to these broad measures. The ideological split between democrats and republicans seems to be a problem from where a start could be made to better support environmentalism. In a same manner, an additional 24.5 percent of explained variance is gained in the evaluative dimension. This is equally due to general values and social positions. With regard to action, not much explanatory power is added by cohorts/gender, general values or social positions because most of their effects are found in the environmental attitudinal variables.

The United States of America is the last country dealt with; the following chapter attends to a summary of the findings of the cross-country macro analysis as well as the micro perspective.

7. Summary of the findings

First and foremost, many of the results are in line with outcomes of previous research. This speaks in favor of the robustness of the findings and the theoretical background. However, not all results are mirrored. Due to the multifaceted approach taken in the present study the issue is more eclectic. It is difficult to summarize such multifarious results and at the same time to find generalizations. Nevertheless, the summary starts with the results of the macro analysis.

7.1 The macro level

Of the three theoretical approaches that explain environmentalism on the macro level – the post-materialism hypothesis, the prosperity and affluence hypothesis, and the globalization hypothesis – the present study supports the prosperity and post-materialism hypothesis with regard to environmental concern and activism. However, many of the relationships are not linear as is suggested by the theories. For the acceptance of laws as well as for everyday action rather the globalization hypothesis prevails. Neither post-materialism nor wealth is related to a nation's acceptance of laws or everyday action. Unfortunately, only 19 countries could be analyzed and these are high-income countries according to World Bank standards. Hence, the results are at best tentative and larger samples including additional, less wealthy nations with good and comparable data quality are required. Nonetheless, the connections revealed should caution future researchers to adopt more careful considerations on the non-linearity of the world.

The macro outcomes also support the idea of virtual environmentalism concerning everyday action. The higher the average level of everyday action in a country, the lower the contribution of environmental concern to the explanation of that action. With activism it turns out to be the other way round. As the level of activism is much lower than the level of everyday action it seems reasonable that in a country where environmental concern more strongly impacts activism, activism is more widespread. It is open to further research to investigate if there is a turning point when a certain amount of activism is reached, and activism increases further, whether the importance of concern decreases again.

With regard to the low-cost hypothesis and attention shifting strategy, the cross-national comparison divides the 19 countries into 4 groups. The first group (Austria, Canada, the Czech Republic, Denmark, Finland, France, Germany East, New Zealand, Slovenia, Spain, Great Britain, Croatia, and the USA) shows the inverted

U-shape in line with the assumptions of Schan (2000; see chapter 3.1.2). The second group (Belgium, Germany West) with the U-shaped curve depicts that for Germany West and Belgium, where every day action is already relatively high, the attention of the environmentally more concerned is shifted to more symbolic acts like supporting laws or activism. Group three (Japan, Norway, and Sweden) resembles the negative linear relationship between an action's cost and the contribution of environmental concern as proposed by the original low-cost hypothesis. The last group comprises only of Switzerland. In this country, the higher the cost of the action, the higher the impact of concern. Even on such a broad level of measurement of environmental concern and action and where only indirect tests of the hypotheses are possible, the low-cost hypothesis and attention shifting strategy pose useful tools for the analysis of people's behavior.

7.2 The micro level

To start the summary of the findings on the micro level, some general remarks about the advantages of the results obtained by the approach taken in this study in contrast to a multilevel method are discussed. The results for the analysis of the cohorts and gender will be summarized afterwards, followed by a summary of the effects of general values and social positions; the last section is devoted to a discussion of the overall contribution of these aspects to the explanation of environmentalism.

The analyses show that there are substantial differences between countries. Country-specific investigations unveiled similarities and differences that a multilevel analysis (taking all 19 countries into account) would have concealed. Take the example of post-materialism: in a multilevel approach the study would have obtained one estimate for the materialists, for example on the conative dimension of concern which is significant. Additionally, when assuming a random slope model, a significant variance of that coefficient would indicate that there is variation between countries. However, in the present study the coefficient in Germany West and the USA is not significant at all. This feature would have remained concealed in a multilevel analysis. From a practical perspective, in a country-specific analysis, a decision-maker could work with the information that in his or her country post-materialist values do or do not play a role in the promotion of environmental protection. It could be misleading if the same person would instead only read that on average, across 19 nations, there is some effect of post-materialism which varies significantly between countries. The assumption of the decision-maker would be to address the less-willing-to-pay materialists to increase their everyday action (because the study indicates that more willingness to pay leads to more action). Unfortunately, if this decision-maker works in the USA where post-materialism does not influence environmental concern, this could result in a waste of money and effort on measures that are unlikely to have any effect. The country profile of the USA, in contrast, communicates that it is not post-materialism but the political world views which are important predictors of environmental concern in the USA.

The scenario described above, of course does not imply that multilevel analysis is not applicable to the analysis of environmental concern. It serves as an example that, depending on the research subject, multilevel analysis might not be the appropriate method despite multilevel data. Multilevel approaches and country-specific analysis should complement each other. Future research could either focus on individual countries as carried out in this study or use a multilevel approach to find out, for instance, which country characteristics (such as GDP or a country's environmental quality) in particular lead to the different sizes of coefficients.

Furthermore, most studies so far only adopt a single measure of environmental concern, in most instances the willingness to pay. This raises the question why environmental concern is discussed as a multi-dimensional construct (see chapter 3.3) when only one dimension is investigated, despite the fact that statistical tools exist that permit the analysis of a multi-dimensional structure. The threefold approach pursued in the present study allows the wide-ranging issue of environmental concern to be assessed more precisely. It shows whether worldviews or positions of people influence how environmental problems are perceived or whether the true differences in environmental concern between people are only unraveled through the evaluation of environmental issues. The multi-dimensional method gives an answer to the question of the constructivism theory, in how far our perceptions and evaluations of environmental issues are prone to subjective social construction; or whether social positions and general values skip perception and evaluation and directly affect behavioral intention or action itself. One-dimensional analyses cannot answer these questions. Overall, the present study yields various new insights and thus adds to the understanding of environmentalism by combining various approaches.

After this overall acknowledgement of the advantages of the present study, 'variable-specific' results will be summarized in the following sections. The predictors of environmentalism are categorized into three groups: the first one includes the aspects gender and birth cohorts. The second group represents general values and finally social positions form the third cluster. The results will be discussed separately for each group. The interpretations refer to the full model so that the interpretations of, for instance, social positions are under control of the variables summarized under cohorts, gender or general values.

7.2.1 Birth cohorts and gender

As discovered in prior research (see chapter 3.2.2.1), the *differences between men and women* concerning their environmentalism are ambiguous in the present study as well. In some countries (the Czech Republic, Norway, New Zealand, USA) women are more environmentally friendly, in other countries (Germany West, Spain) women are more concerned regarding certain aspects but less concerned in another. Only in Germany East women are less environmentally friendly than men. This is indicated by the significant negative effect on the evaluative dimension and by the non-significant ones on the other aspects. Independent of general values or social positions, the way in which men and women think and act in relation to the environment is moderated by country specific elements, which are not captured in this study. Gelissen (2007) and Swamii et al. (2011) state that it is not gender per se but gender-specific socialization, "a motherhood mentality or an ethic of care" (Gelissen 2007, p. 399), as well as gender-related personality traits that mediate the gender effects on environmentalism. The overall outcome of this mediation is moderated by macro aspects as indicated by the country -specific results.

The results for the *birth cohorts* are mixed as well, but a general tendency can be identified: people born between 1964 and 1983 are most environmentally concerned. The youngest cohort, i.e. born 1984 or later, act in the least environmentally friendly way, rather object to laws, but are_more involved in activism. The first trend can be explained by the increasing importance of the environment within the public and political agenda in the 1960s and 70s. On taking a closer look, the tendency of an inverted U-shaped curve (Blaikie 1992) emerges, most evident in the evaluative dimension (Figure 14).[23] To some extent in all countries the effects become stronger the later a person was born and at some point they become weaker again. The clearest pattern is found in Norway, Spain and Germany West while in the Czech Republic there are barely any differences between the

23 Note that in Figure 14 the unstandardized coefficients of the full model before removing the non-significant paths are shown. This was done to have a coefficient for each cohort to exemplify the trajectory. When the significant coefficients are removed, only the significant peaks would appear in the figure.

cohorts.[24] Similar patterns also exist for the cognitive component of concern (e.g. in New Zealand and Norway). The second trend, i.e. that younger people are less active, has already been shown by other studies (Engel and Pötschke 1998, Hadler and Haller 2011, Olofsson and Öhman 2006, Olsson 1994, Preisendörfer 1996b, Swami *et al.* 2011). The explanation for this development is more suggestive. Some of the youngest cohorts may still live at home with their parents where they have less control over these actions or simply act carelessly regarding behaviors such as saving water, energy or buying environmentally friendly products. In contrast to everyday action or the acceptance of laws, environmental activism may be a more 'appropriate' activity for young people and therefore their attention is shifted toward this conduct.

Figure 14: Cohort effects on the evaluative dimension of environmental concern

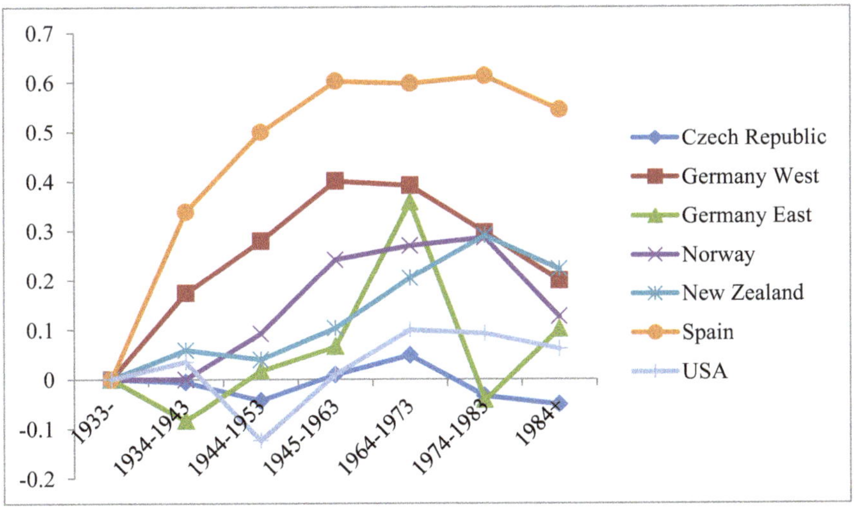

Despite these two general trends, there are still substantial country disparities. For detailed country differences the respective profile should be consulted. Particularly noteworthy are the effects in New Zealand and Norway. The same cohorts that exhibit higher values in the cognitive or evaluative dimension are at the same time less willing to act or act less frequently. Another example would be the higher willingness to pay among younger Czech people which are at the same time less active.

24 As can be seen in Table CZ 7 none of the cohort's coefficients is significant represented by the blank cells in the table.

In the USA, in contrary, positive effects from the cohorts emerge only on the cognitive and evaluative dimension while the conative element and action are unaffected except for on significant effect: the cohort born between 1934 and 1943 acts in a slightly more environmentally friendly way in everyday life than the others.

Although the cohort effects revealed some important insights from a practical perspective further research is suggested to delve deeper into that issue. The present results cannot determine whether the effects found are the product of cohort-specific socialization processes or a snapshot of life-cycle effects. Country-specific features of the period, including key characteristics of the time people spent their childhood or adolescences in, need to be taken into account. To investigate the life cycle effects that are mentioned in the theoretical part (see 3.2.2.1) panel data is necessary that is able to trace the individual's development of environmentalism. From the findings in the present cross-sectional analysis it can at best be concluded that people in their thirties and forties are most environmentally concerned (with country-specific variation) and that the most ambiguity prevails among the youngest. This may be an indication for a life-cycle effect.

7.2.2 General values

At this point it is important to recall, that in the present study general values are conceptualized in a broad way. All three measures of general values (post-materialism, political ideology, and trust) are to be understood as proxies for the concrete corresponding values that can be summarized under each of the three concepts as no detailed variables are available. Despite this broad conceptualization of general values the results show that they contribute to the explanation of environmentalism, although to varying degrees depending on the country as well as the aspect of environmentalism under the microscope.

Most straightforward are the results of *post-materialism*. In contrast to the assumption that post-materialism is an idealistic perspective on the environment and should only affect the idealistic parts of environmentalism (see 3.2.2.3), effects were found on all three dimensions of environmental concern as well as on all three types of action. Only in the USA are materialists and mixed types equally as environmentally concerned as post-materialists. The strongest impact is observed in New Zealand and Norway, whereas a rather weak influence is found in the Czech Republic and Germany East. The most pronounced effect of post-materialism on action is found on activism (direct and indirect effects) while for everyday action mostly mediated effects via concern emerge. Even though environmental pollution and problems are very material issues, the cognition and evaluation of environmental issues as well as the reaction to them are still matters of ideals, at least in part as the only small-to-moderate effects show. Surprisingly, in the Czech Republic, New Zealand, Spain and the USA there is a higher support for laws among the materialists than for mixed types or post-materialists (concerning direct effects). As mentioned in the profile of the Czech Republic, several mechanisms might be behind this – at first sight unexpected – relationship. The material aspect of environmental problems (e.g. a threat to the individual's health, economic burdens, etc.) increases the support for laws among materialists in several parts of the world. Moreover, due to their lower concern in general, materialists in these countries perceive the responsibility of action to be the government's task and less the responsibility of individuals on a voluntary basis. The post-materialist's emphasis on self-fulfillment or political freedom could also be an explanation why in these countries the materialists' support for laws is stronger than the support of mixed types or post-materialists. Laws might be perceived as restrictions to self-fulfillment or political freedom and are therefore less supported by post-materialists.

The *political position* yields the least clear results, yet there is a discernible tendency that environmentalism is present on the left side of the political spectrum. This relationship is not necessarily linear and there are differences between countries. For example, in Germany West only the left voters are more concerned while in the USA the political position's impact on the cognitive dimension of concern is close to linearity. In contrast, in the Czech Republic the central and right/conservative voters are most willing to pay and are also more likely activists.

The following paragraph will provide some explanations for these country differences. The categorization of the political positions is based on peoples' political party preferences. These preferences' influence on environmental concern, as stated before (see chapter 3.2.2.3), is "a function of country-level characteristics" (Nawrotzki 2012, p. 286). These characteristics, such as environmental conditions, the dependency of the economy on the environment (e.g. ecotourism) or a communist history, may play a different role in the respective political parties in different parts of the world. For instance, in Germany it was a center right/conservative coalition that executed the energy transition, although with the support of Social Democrats and Greens. Nevertheless, this coalition supported nuclear power before the accident of Fukushima. Thus the country specific variation is only surprising at first sight but when taking into account the theoretical considerations, the complexity of environmental problems (see 'An excursion: environmental pollution') as well as differences in the importance of particular environmental problems, the variations make sense. A complementing argument explaining the inconsistency of the effects of the political position, in particular where it loses importance, may be borrowed from the globalization hypothesis. The hypothesis states that environmentalism spreads across the world, independently of wealth. This spreading might also include spreading across different social groups, a so-called diffusion of innovation (Rogers 2003). In diverse societies, environmental concern is first adopted by one group, e.g. the political left. Over time it diffuses into society and is assumed by its groups at varying paces. Thus right/conservative voters (as in the case of New Zealand) can judge environmental problems to be not as dangerous as expected based on theory and yet show more evaluative concern, which one would not expect intuitively.[25] The negative effect on the cognitive dimension and the positive effect on the evaluative element of environmental concern might be explained in the following manner: 'environmental problems are not that bad (now), and, if we care for the environment, they will not become severe burdens'.

Leaving the complex issue of political world views aside leads to a less complicated value, to *trust*. The measure of trust in this study is a measure of generalized trust. The results for trust are quite clear and the tendency of its effects is the same for all country samples. In most cases higher trust leads to higher environmental concern with regard to the evaluative and conative dimension, and to a less intense perception of environmental problems. It can be said that trust can be an aid to overcoming people's unwillingness to do something or their skeptical positions toward environmental issues. The negative impact on the cognition of environmental threats and the thus reduced indirect and total effects on action, admittedly, are striking. When thinking about the causal mechanisms behind the impact of trust one might think of trustful people considering environmental problems as less serious, as more trustful people tend to look at the world less pessimistically. This argument would also be in line with the positive effect of trust on the evaluation and the intention to do something. Less pessimism and skepticism created by trust increases evaluative concern and the willingness to act because it is trusted, others will do so as well. With regard to action, it is predominantly the acceptance of laws and activism that are supported by trust. In these two aspects the evaluative dimension of concern is most important and for everyday action the cognitive dimension is the stronger predictor. Consequently, more of the positive effects of trust on the evaluative dimension are transferred over to laws and activism. Likewise, more of the negative impact on the cognitive dimension is mediated on everyday action. Generally, trust is more influential

25 Note that the positive effect of the right/conservatives disappeared once controlling for social positions.

on the value-rational element of environmentalism than on the instrumental-rational element. For the integration of people when considering how to deal with environmental problems, trust, as a key feature for a functioning democracy (see chapter 3.2.2.3), plays a non-negligible role.

7.2.3 Social positions

As was the case with general values, the social positions captured in this study are broad measures. Especially the *occupational position*, the number of *children and toddlers* in the household and *community size* do not seem to be suitable factors for explaining environmental attitudes or action. Although in many instances some coefficients are significant, their size mostly lacks practical relevance, and significance might only have been achieved due to the large sample sizes. Country-specific in-depth analysis with detailed measures appears more promising to get a better insight into the meaning of these positions than the broad conceptualizations used in this study. Only the occupational positions have effects with some practical relevance – judged relevant due to the $b>0.1$ – predominantly on activism and the acceptance of laws (in the Czech Republic, Germany West, Norway, the USA). A more stable occupational situation seems to augment environmentalism at least in some contexts.

Education, in contrast, is the most stable predictor of environmental concern in all six samples. Conversely, education's effects on everyday pro-environmental action are less pronounced and less consistent. In the USA people with a university degree are more often engaged in such activities while in the remaining countries only the two parts of Germany have significant direct coefficients. In Germany East this kind of action is less widespread among the better educated for reasons suggested in the profile. In Germany West, every educational group is more active than those with no formal education. The best educated, however, to are active to a lesser extent than those with a lower or intermediate degree. In part this lower direct participation of the better educated in action is outweighed by the indirect effects higher education has on action via the evaluative dimension of concern. Across the six countries activism and the acceptance of laws, in general, are higher among the better educated.

These manifold relations between education and environmentalism as well as the different slopes underline, on the one hand, that the grade of education is important when people are considered as key actors in the solution of environmental issues. On the other hand, it raises the question why the effects are not large? Education not only co-varies with environmentalism but also with many of the other predictor variables, inter alia trust or income. This indicates that education is a complex issue itself, barely comparable across cultural borders. It could even be suggested that within one nation some of the effects of education might accrue due to differences in the type or area of education. Imagine people with a university degree, for example. Did they study environmental sciences and are thus by nature of the subject more environmentally concerned? Or was their major subject business administration at a university where environmental management or environmental economics are not part of the curriculum? Are conservative voters who are well educated less concerned than green party voters with the same degree of education? Are the effects of education moderated by differing world views? These questions cannot be answered within the framework of this study and are open to future research. Despite these unresolved riddles, the outcomes for education point in one direction: education increases environmentalism. Consequently, it is the educational system in which environmental issues need to be included from a practical perspective in order to intensify environmentalism. However, it has to be kept in mind that education is a long-term process. To achieve a university degree in Germany people have to attend school for 12–13 years followed by a minimum of three years for a bachelor's degree and two more years for a master's degree. In the best-case scenario, a master's graduate spends 17 years in the educational system. Better integration of environmental subjects into the educational system will have its effect only in the long run.

Turning now to *income*, on the micro level, the affluence hypotheses would not find full support regarding the six countries analyzed. The effects of the economic position within the country, represented by the household equivalence income, are multifaceted. Three groups of countries appear. In the Czech Republic and Germany West the effect follows the expected direction. In both countries wealthier people are more environmentally concerned. In Germany a higher concern is exhibited in the evaluative dimension. In the Czech Republic wealthier people are more willing to pay. In contrast, in Germany East higher income has a substantial negative impact on the cognition of environmental threats and in Norway on the evaluation of environmental issues. The third group comprises New Zealand, Spain and the USA, where a higher equivalence income decreases the perception of threats and at the same time it increases the evaluation (in Spain also the intention to do something). As theory and research on environmental justice suggest, more wealthy people are often less exposed to pollution or environmental problems (e.g. Bell 2012, pp. 25–33, Diekmann and Meyer 2010, Mohai and Saha 2006, Schlosberg 2007, Shepard *et al.* 2002). Hence, they perceive environmental threats as less dangerous. The positive coefficients on the other dimensions are in line with the reasoning of the wealth and affluence hypothesis.

Focusing on environmental action, there is no single direct effect of income on everyday action. The acceptance of laws is higher among the wealthy people in Germany West and lower in Norway while it has no impact on the acceptance of laws in the other countries. On activism, only in the USA does a direct positive income effect emerge.

From a scientific perspective, future research should focus on the concrete socio-economic situation that creates the ambiguous wealth effects on environmentalism. From a practical perspective, one would have to address less wealthy people on areas where they are less concerned in two ways. First, their own environmentalism can help reduce their own environmental or economic (saving energy saves money) burden. Secondly, policy makers have to stabilize or improve their economic situation so that environmental issues have the possibility to grow in importance. Luckily, stabilizing or improving people's income situation appears to be some kind of universal goal for policy-makers so that this issue is already on the agenda. Finally, there are the cases where a higher income reduces environmental concern. There is a need to find the mechanisms behind this relationship so that their less environmentally friendly position can be better included when developing measures that build on support from the population or that are intended to improve people's pro-environmental action.

7.2.4 The overall contribution of birth cohorts, gender, general values, and social positions to the explanation of environmentalism

Environmentalism is a multifaceted topic and the findings throughout the course of this study underline this statement. The hypothesized predictors do not consistently contribute to environmentalism across the six countries. The share they have in the explanation of environmentalism also varies between the six nations as Figures 15a-f depict. The road from perception to reaction is rocky and many more individual factors guide the way in addition to what is found in this study. In the face of the relatively high number of predictors already included, the diffusion of environmentalism into society has by now advanced relatively far. As a result the impact of concern and the other predictors is especially low where the level of action is comparably high. This supports the idea of virtual environmentalism (see 3.2.2.1), making environmentalism easy and thus reducing the importance of environmental concern, values or positional factors. Yet, in different countries the three groups of predictors contribute to different extents (see Figure 15).

Figure 15: Explained variance of environmental action and concern by model

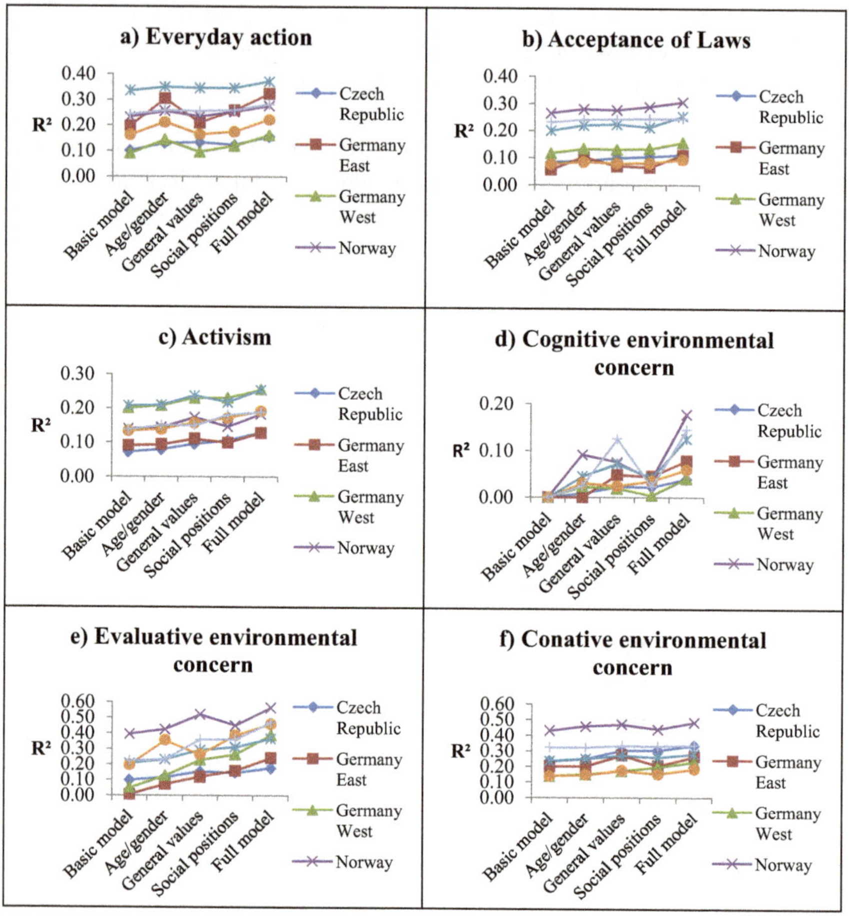

Starting with *everyday action* (Figure 15a), comparing the explained variance in the basic model that includes only the three dimensions of environmental concern and action to the full model shows an increase by only 4.0 percentage points in the USA to 12.7 points in Germany East. On the one hand this indicates that there are additional explanatory variables behind environmentalism than accounted for in the present study. "In short, environmentalism is most likely a complex social phenomenon, a mixture of social perceptions, local histories and environmental realities, international relationships and influences, and unique cultural and structural features of particular countries and regions" (Brechin 1999, p. 807). This statement implies that "attitudes are forever changing naturally" (Heberlein 2012, p. 147), because the aspects mentioned by Brechin change as well.

Regarding everyday activities, in the Czech Republic, general values matter most. In Germany, New Zealand, Norway, and Spain, cohorts and gender have the strongest effect and in the USA, social positions increase the explained variance most. The numbers, in most cases, are quite close to each other so it would be exaggerated to speak of substantial differences.

Regarding everyday activities, in the Czech Republic general values matter most. In Germany, New Zealand, Norway, and Spain, cohorts and gender have the strongest effect and in the USA, social positions increase the explained variance most. The numbers, in most cases, are quite close to each other so it would be exaggerated to speak of substantial differences.

In a similar manner, the *acceptance of laws* (Figure 15b) and *activism* (Figure 15c) experience little growth in the explanation of their variation when predictors are added. The cohorts and gender, general values and social positions explain the acceptance of laws equally badly, with the exception of Germany East where belonging to a certain birth cohort and gender stands out: in the eastern part of Germany the cohorts and gender have a discernibly greater impact on the acceptance of laws than the other aspects. With regard to activism, from one model to the other, the explained variance increases slightly with a small break among the social positions in Germany East, New Zealand, and Norway. The country-specific effects of environmental concern on all three types of action (basic model) can be found in tables 4.2–4.4.

Moving from action to environmental attitudes, in general the *perception of environmental threats as dangerous* seems to be a widely diffused phenomenon. This is reflected in the relatively high level (Table 4.1) and in the small share that the cohorts, gender, general values and social positions contribute to its formation (e.g. low R^2, Figure 15d). The greatest impact of the cohorts, gender, general values and social positions is present where also the largest effect of environmental concern on action was found: in New Zealand, Norway and the USA. However, even in these countries R^2 is lower than 0.2. Concluding, most people agree that environmental threats are, to some extent, dangerous. Despite the widespread preference of researchers for large coefficients and high explained variances, from a practical point of view this result is promising: if there are fewer objections to the question of whether environmental problems pose a risk or not, communication and implementation of environmental protection measures might be far easier; and less effort is needed to develop strategies that help overcome disparities between occupants of different social position or people that hold conflicting values.

Yet, the three groups of predictors have at least some explanatory power that must not be disregarded. Figure 15d unveils which of them has the largest share in the formation of the perception of environmental issues. In most countries, general values contribute strongest (the Czech Republic, Germany East, New Zealand and the USA). In Germany West and Norway the cohorts and gender play the most important role while only in Spain are social positions most influential.

When recalling the complexity of environmental pollution and problems from the excursus at the beginning, in the real world the perception of environmental threats is an intricate matter. Despite the overall high cognitive concern, represented by rating items, the country-specific perception of the most important environmental problems should be taken into account (ranking items, as presented at the beginning of each profile). Thus, decision makers are able to find the most pressing issue where environmental protection can start.

In addition to this general perception, there is people's *affective and normative evaluation* of environmental issues as well. If a problem is perceived as dangerous but people do not think they can or should do something about it or there are more pressing issues, the mere perception might not suffice to evoke action. In some countries the cognitive dimension influences everyday action directly (e.g. New Zealand), but in others only indirectly via the evaluative dimension of environmental concern (e.g. Germany West). It depends on the individual country whether the mere perception of a threat leads to commitment in dealing with the threat or if there are mechanisms in between (such as the normative evaluation) that are at work before turning into action. If this mediation is true, it is even more important to think about the general values and social positions that influence the evaluation of environmental issues. In the evaluative component of environmental concern, the growth of the R^2 is largest with the addition of further predictors (Figure 15e). For instance, in Germany West, the evaluative dimension is the only direct predictor of environmental action. This dimension is influenced most by social positions and in the full model the explained variance is 33.8 percentage points higher than in the basic model (Figure 15e). It starts with 4.7% explained variance by cognitive environmental concern and ends with 38.5 percent in the full model. In contrast, in the Czech Republic merely an additional 7.6 percentage points are gained in the full model.

As regards the cognitive dimension, it depends on the country whether the cohorts and gender, general values, or social position add most to the explanation of evaluative concern. In the Czech Republic and Norway general values contribute most, but to a larger extent in Norway than in the Czech Republic. For the other nations it is different: in both parts of Germany, New Zealand, Spain and the USA the effects of social positions are strongest. In countries where general values mattered most, social positions were of second-highest weighting. In total, the evaluation of environmental issues 2010 is a question of subjective construction represented by the effects of general values and social positions.

Figure 16: Effects of the cognitive and evaluative dimension of concern on the conative dimension by country

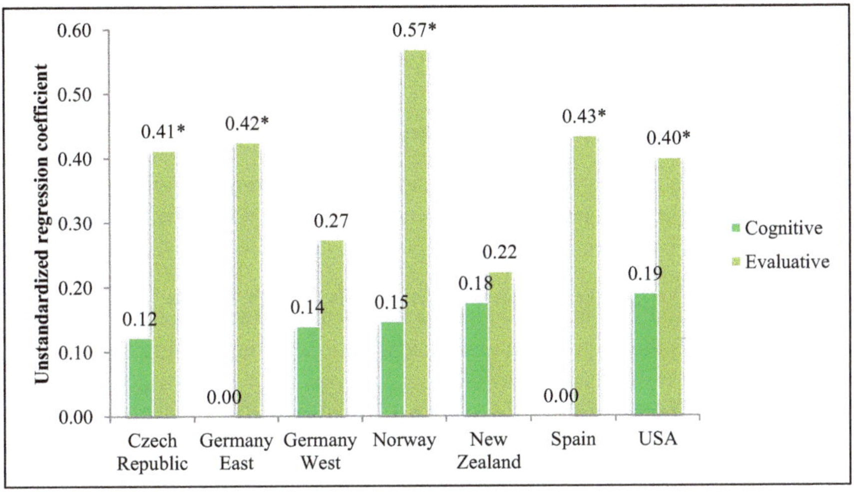

Notes: * statistically significant difference between the coefficients of the cognitive and evaluative dimension; unstandardized coefficients

Furthermore, in most countries, some part of the effects of the cognitive or evaluative dimension on action is mediated via the *behavioral intention*, i.e. the people's *willingness to pay*. In this aspect of environmental concern, the explained variances are relatively high, but as Figure 15f depicts, the direct contribution of cohorts and gender, general values, or social positions are marginal. With an R^2 increase of 0.102 the impact is largest in the Czech Republic, and the additional 1.5 percentage points increase in the explained variance in the USA indicate a virtually absent relationship of the conative environmental concern with cohorts/gender, general values and social positions. In the Czech Republic general values and social position contribute equally, in Germany West social positions are most important while in the remaining countries general values matter most. Of the overall explained variance of the conative dimension, most is attributed to the evaluative and cognitive dimension of concern. Therefore, in addition to showing which of the group of predictors is most important it seems reasonable to address which component of environmental concern increases the behavioral intention most (Figure 16). In all seven samples the evaluative dimension exerts a stronger influence on the intention to act than the perception of threats does. In Germany East and Spain the coefficient of the cognitive dimension on the willingness to pay is not significant at all, as represented by 0.00 in Figure 16. In all countries the behavioral intention is value-rationally rather than instrumental-rationally driven. Note that the coefficients of the cognitive and evaluative dimensions on the conative component in Germany West and New Zealand are not significantly different from one another.

7.3 Limitations of the present study

After the lengthy discussion about the broader framework in which environmental action is embedded, the elaboration of the theoretical background and finally the long chapters aiming at the explanation of environmentalism, another issue needs to be addressed: the limitations of this study.

First of all the ISSP only provides data on self-reported behavior. Perceived normative expectations or social desirability thus can influence the respondents answering behavior (Blasius and Thiessen 2012, pp. 6–8). There might be some degree of over- or underestimation of one's own activities. However, Skarbek-Kozietulska et al. (2012) found that about two thirds of respondents who the authors knew had committed one or more minor delinquent acts responded truthfully in a survey. Looking at the descriptive statistics in Table 4.1 the frequency of environmentally friendly answers declines the closer one approaches action. If the responses provided in the ISSP were unduly a function of social desirability higher frequencies of action would be expected. These previous findings and the analysis of data quality in chapter 4.2 support the assumption that the researcher can rely on the data provided; and that meaningful analysis with self-reported behavior is possible.

Second, only a very narrow area of everyday activities is captured by the three analyzed items (reducing energy use, saving water and avoiding certain products). These are low-cost or even money-saving activities. By choosing these three activities from the questionnaire it was assured that the activities that are analyzed are available for everyone. Respondents were not able, for instance, to conceal their non-action by stating saving water or energy is not available. In contrast, the relatively high number of missing values indicating non-availability prevented the other three items covering different activities from being used in the present study. Recycling behavior depends on a recycling program, buying untreated fruits and vegetables requires shops where such products can be purchased, and cutting back car driving is only an option if a car is accessible.

Another issue is the sample size. The small sample of Germany East in particular poses the risk of a lack of statistical power to achieve good estimates. Yet, the confidence intervals are of comparable size to those of the other countries, so the estimation can be considered equally precise. Nevertheless, larger samples would be desirable. This allows more precise estimates especially if the models become more complex. Likewise, larger samples entail that small subgroups (e.g. people with no educational degree) become large enough to be included in the analysis as a separate group. In most countries these had to be combined with the subgroup including those with the lowest formal education due to the fact that less than ten respondents stated they had no formal education.

Furthermore, the present study provides a country-level understanding. As the separation into Germany East and West already reveals, micro analysis on a regional or state level seems promising in order to get more detailed insights. It is plausible to consider regional differences also in other countries. To allow analysis, for example, on federal state level, larger sample sizes would be necessary. It remains, however, doubtful that such large samples are likely or to be collected or if this is feasible. An immense amount of time and money would be required to achieve a sufficient number of completed questionnaires per country and region.

Finally, as in the nature of secondary analyses, the scholar is restricted to the items of the questionnaire decided upon by the initiators of the survey. Consequently, the user, to some degree, has to adapt the research question and selection of appropriate variables to available resources. This is a tradeoff many researchers have to accept when they want to conduct cross-national analyses. Prior to substantive analysis the researcher has to familiarize himself with the data and a rigorous assessment of the data quality is inevitable. If this is done and the questionnaire covers the most important facets of the topic under scrutiny, important insights can be gained by secondary analysis.

8. Concluding remarks

"When it does not take an act of Congress for people to significantly act upon climate change, then perhaps we will have begun to turn the tide" (Lejano *et al.* 2013, p. 69). In the same vein, the overall question of the present study is how to better get people involved in the protection of the environment, for themselves and future generations, not as a mere function of parliament enactments but as proactive socio-economic, political and legal actors. Therefore, first and foremost, some practical conclusions can also be drawn from the study at hand.

The descriptive inspection of the salience of environmental issues shows that there are some country particularities. The country-specific features need to be incorporated into environmental policies when public actors try to include people's point of view on environmental topics. They need to build on the attitudes and should not work against them (Heberlein 2012, pp. 140–148). Is there a disparity between objective environmental problems and their subjective perception? Or are there objective problems that also are perceived to be a problem? Taking such and similar questions into account will help a decision-maker to know which problem to address first. To give an example, the analysis underlines that politicians or power companies can count on broad support for renewable energies in the population on which they can built energy policies or management strategies. Likewise, climate change is considered the second-most important environmental problem across the six countries, with country-specific variation, and again, decision makers should establish measures that incorporate the concern about climate change, taking cross-country similarities and differences into account.

Throughout the global profile, the country profiles and the summary of the findings, notes and comments on the practical relevance of various results were provided. Education, post-materialist values and trust are the three universal predictors of environmentalism – mainly of environmental concern – and therefore the focus of this conclusion is placed on these three aspects. Trust and education offer the most promising starting point from a practical perspective. Both have a hand in every democratic and civil society. To paraphrase Niklas Luhmann (1979, p. 4): without trust we maybe would not get up in the morning or leave home and participate in everyday life as nearly nothing we do in modern societies is absolutely independent of the behavior of others. Trust, in other people as well as in the government and societal institutions, e.g. politics, the economy or the legal system, is de rigueur for a functional and sustainable democracy as well as for the fostering of cooperation (Blind 2006, Jalava 2003, Misztal 2001, Walter 2010) and to help people take part in a multiple stakeholder approach (see below). As the present study shows, trust is also important to integrate people in environmental protection. Building up people's trust, e.g. by demonstrating that indeed something has been done or can be done rather than only promising or pretending to do something seems crucial for the solution of environmental challenges. The first step should be to take the lead as a good 'role model'.

The second-most stable and consistent predictor of environmental concern, education, is a point where decision- or policy-makers can start addressing people as socio-economic, legal and political actors at the institutional level: the educational system. Not only natural science education about biological, chemical or physical procedures is important to provide a better technical understanding of the environment. Health education should also include environmental health threats. Social, legal as well as economic education must not neglect the environmental aspects with which these areas are interwoven, as described in chapters 1 and 2.

One causal mechanism behind education's effect on environmentalism is that a higher education entails the capability to better consider, reflect and critically evaluate a complex issue such as the environment. This may well be transferred to other matters. Consequently, improving education means improving a universal competency to deal with all the intricacies of the world we live in. So, education also matters for the individual itself. Better education widens job prospects and usually offers a higher income. Education is a means of accumulation of human capital which "implies the acquisition of knowledge and skills that permit an individual greater efficacy in navigating through life" (Dietz *et al.* 2009, p. 118).

From a macro perspective, a better-trained workforce allows an economy to fill empty spots that technological innovation creates more quickly with appropriate personnel. Likewise, jobs created through ecological modernization often also need qualified specialists (see chapter 2.1). Similarly, with better education, abstract and complicated laws would be easier to understand (see chapter 2.2) and the claims or statements of environmental protection groups can be better reflected upon and comprehended, which is crucial for the selection of the appropriate group or form of activism (see chapter 2.3). After this digression on other positive 'side effects' of education, it is clear that education is a key aspect in the solution of societal problems, *including* environmental problems.

Yet, there is a need for a final word on education. Speaking about education in the context of this study means formal education and the time people spend in the educational system. The effects education has on environmentalism are the result of a long process and should not be confused with short-term education or information. The chapter "Educating the public … and other disasters" as Heberlein (2012, p. 69ff) provocatively calls it, shows that simply providing information on environmental issues and educating the public in that way has a very limited scope for changing people's behavior. For an elaboration on the issue of short-term 'educating the public' the reader is referred to Thomas Heberlein (2012: chapter 5).

Both the fostering of the people's trust in each other and politics as well as an educational development are medium- to long-term issues that do not develop from one day to the next. For that reason, two aspects of the practical and theoretical perspective of this study need to be recalled. In chapter 2.1 it was already mentioned that in the long run the conflicts between economy and the environment disappear and in 3.1.2, when elaborating the subjective rationality perspective, it was said that short-term rationality often predominates the long-term perspective. Consequently, this study supports the urgent need to expand thinking to a long-term perspective, keeping in mind that a measure taken now takes some time before coming into effect, with the costs usually occurring at once. An investment in education or the increase of trust is an investment in the future.

In contrast to education and the educational system, the materialism-post-materialism dimension is not institutionally embedded. It is a set of values people hold. First of all, despite the quite narrow measures of post-materialism in the ISSP, it generally captures a broad spectrum of values. Secondly, in democratic societies, where freedom of speech and opinion are crucial, dealing with or changing people's values is a highly delicate matter. Therefore, instead of trying such a value change, decision-makers should communicate environmental issues in an appropriate way. It should be clear that these problems are not just a matter of idealism (implied by post-materialists' world views, see also 3.2.1). They have very direct implications on the material side, for instance for price stability (e.g. rising prices of food due to water shortages) or social security (e.g. resource conflicts or migration). Issues like environmental pollution and problems that have real material consequences need to be dissociated from post-materialism or idealism.

In this spirit, the central task of the present study is to provide an understanding of how to open doors and guide the way for people in their role as socio-economic, legal and political actors in relation to the environment. It shows, inter alia, which general values or social positional factors increase or decrease people's concern, behavioral intention or action itself. A very important finding is the support for the assumption of virtual environmentalism. When environmentalism is made easy through good socio- and infrastructural conditions, behaving in an environmentally friendly manner is less dependent on social positions, general values and attitudes. This leads to a diffusion of environmentalism into the population. This, however, complicates the addressing of those who still show deficits concerning their environmentalism as these deficits are less tied to concrete characteristics. The social-structural conditions for an easy environmentalism need to be constructed in a way that is consistent with people's values and attitudes, be they more or less environmentally concerned: decision-makers need to go with the flow and not try to swim upstream as Heberlein (2012: chapter 9) would illustrate.

After these general practical implications, a policy approach called neo-corporatism that includes people as crucial actors will be outlined as the last part of this study. With reference to chapters 1 and 2, the integration of the public into policy, legal, social and economic thinking is crucial and also promising as illustrated by the following; the present study shows several aspects that could serve as a starting point in this respect for decision-makers or people already committed.

What does corporatism actually mean? Corporatism is an idea where various actors participate in public policy and governance; it is a multiple stakeholder approach. Generally speaking, sustainability governance policies concerning one area (e.g. social, economic, health or ecologic governance) must not be implemented at the cost of other areas (Foley 2004, Walter 2010). These goals have to "be met in an integrated way" (Walter 2010, p. 133) to gain the benefits all areas offer and at the same time to minimize the costs, bads or risks. In fact, all of these areas do not simply coexist, they are interdependent. As it is the leitmotif of the whole study, people are the actors within these interdependencies in their day-to-day routine, in their support for jointly negotiated legal measures or as activists (see chapter 2). They are one important actor among the multiple stakeholders.

Some scholars discover positive effects of corporatism on environmental governance, i.e. by integrating science, economic stakeholders, and the public into policy making on different environmental issues (Crepaz 1995, Jahn 1998, Matthews 2001, Scruggs 1999, 2001). Only little evidence is found that does not discover positive effects of corporatism (Neumayer 2003). For an overview, see Stauffacher (2010).

In the context of environmental governance, Staufacher (2010) implies modernizing the idea of corporatism by what is called neo-corporatism. Neo-corporatism "stands for more negotiated forms of policy making" (Stauffacher 2010, p. 202). From a modern and transdiciplinary perspective, the integration of science, economic stakeholders and the public are essential. Neo-corporatism, which like corporatism is a multiple-stakeholder approach, is rather seen "as mutual learning process than as interest negotiation" (Stauffacher 2010, p. 203). Networks between different groups need to be developed that go beyond hierarchy and market rules (Traxler *et al.* 2001, p. 301). This allows social and ecological issues to remain an important matter on the agenda without putting other problems aside. Unintended consequences in a Mertonian manner (Merton 1936) can thus be reduced or even avoided at all. The neo-corporatism as a multiple-stakeholder approach was found to yield results that are economically neutral and less harmful from a social and ecological perspective (Bornschier 2005). Success of such an approach can only be achieved with the inclusion of the public as socio-economic, legal and political actors.

Neo-corporatism offers a practical way to elaborate good social structural conditions, especially with the inclusion of the public as legal and political actors. The socio-economic actor, in this study, is represented by the individual's activities in their day-to-day routine which at first sight might seem unlikely to be part of corporatist approaches. However, being a socio-economic actor also means networking, communication, interaction and the exchange of ideas. When neo-corporatism provides opportunities also for the socio-economic actor within a community, it will also support people in their everyday activities in learning from others, by suggesting own ideas and jointly developing new ones that can be incorporated into the daily routine.

Finally, this study began with an excursus on the multi-facetedness of environmental problems as well as their human causes and thus the circle is complete: environmental problems are not only complex matters often caused by human activities. There is also still much to do for a long-term sustainable society; and indeed a lot can be done by human activities. As pointed out in chapters 1–2, people are crucial actors in dealing with present and upcoming environmental pollution, degradation and problems. Nevertheless, people need to be 'activated' to play this part: either as socio-economic, legal and political actors within a neo-corporatist approach or by another tactic, as this study shows, with the example of six countries where politicians, decision-makers or environmental groups will find a starting-point to activate environmentalism.

Literature

Abramson, P. and Inglehart, R.F., 1995. *Value Change in Global Perspective*. Ann Arbor: University of Michigan Press.

Adeola, F.O., 2000. Cross-National Environmental Injustice and Human Rights Issues A Review of Evidence in the Developing World. *American Behavioral Scientist*, 43 (4), 686–706.

Akella, A.S. and Cannon, J.B., 2009. Strengthening the Weakest Links: Strategies for improving the Enforcement of Environmental Laws Globally. *In*: R. White, ed. *Environmental Crime: A Reader*. Portland: Willian Publishing, 528–567.

Alcott, B., 2005. Jevons' paradox. *Ecological Economics*, 54 (1), 9–21.

Apel, D., 1992. *Verkehrskonzepte in europäischen Städten: Erfahrungen mit Strategien zur Beeinflussung der Verkehrsmittelwahl [Traffic Concepts in European Cities]*. Berlin: Deutsches Institut für Urbanistik.

Atapattu, S.A., 2006. *Emerging Principles of International Environmental Law*. Ardsley: Transnational Publishers.

Attari, S.Z., DeKay, M.L., Davidson, C.I., and Bruin, W.B. de, 2010. Public perceptions of energy consumption and savings. *Proceedings of the National Academy of Sciences*, 107, 201001509.

Bamberg, S. and Möser, G., 2007. Twenty years after Hines, Hungerford, and Tomera: A new meta-analysis of psycho-social determinants of pro-environmental behaviour. *Journal of Environmental Psychology*, 27 (1), 14–25.

Barkin, J.S., 2003. Realist Constructivism. *International Studies Review*, 5 (3), 325–342.

Barnosky, A.D., Hadly, E.A., Bascompte, J., Berlow, E.L., Brown, J.H., Fortelius, M., Getz, W.M., Harte, J., Hastings, A., Marquet, P.A., Martinez, N.D., Mooers, A., Roopnarine, P., Vermeij, G., Williams, J.W., Gillespie, R., Kitzes, J., Marshall, C., Matzke, N., Mindell, D.P., Revilla, E., and Smith, A.B., 2012. Approaching a state shift in Earth/'s biosphere. *Nature*, 486 (7401), 52–58.

Barr, S., 2007. Factors Influencing Environmental Attitudes and Behaviors A U.K. Case Study of Household Waste Management. *Environment and Behavior*, 39 (4), 435–473.

Barr, S., Gilg, A., and Shaw, G., 2011. Citizens, consumers and sustainability: (Re)Framing environmental practice in an age of climate change. *Global Environmental Change*, 21 (4), 1224–1233.

Baumol, W.J., Oates, W.E., and Blackman, S.A.B., 1979. *Economics, environmental policy, and the quality of life*. Prentice-Hall.

Beall, J., Guha-Khasnobis, B., and Knabur, R., 2010. Beyond the Tipping Point: A Multidisciplinary Perspective on Urbanization and Development. *In*: J. Beall, B. Guha-Khasnobis, and R. Knabur, eds. *Urbanization and Development: Multidisciplinary Perspectives*. Oxford: Oxford University Press, 3–16.

Beck, U., 1992. *Risk Society: Towards a New Modernity*. London: SAGE.

Beck, U., 2010. Climate for Change, or How to Create a Green Modernity? *Theory, Culture & Society*, 27 (2–3), 254–266.

Bell, J.N.B., Honour, S.L., and Power, S.A., 2011. Effects of vehicle exhaust emissions on urban wild plant species. *Environmental Pollution*, 159 (8–9), 1984–1990.

Bell, M.M., 2009. *An Invitation to Environmental Sociology*. Los Angeles: Pine Forge.

Bell, M.M., 2012. *An Invitation to Environmental Sociology*. Los Angeles: Pine Forge.

Van den Bergh, J.C.J.M., 2007. Abolishing GDP. *Tinbergen Institute Discussion Paper*, 19 (2).

Bergmann, M. and Klages, M., 2012. Increase of litter at the Arctic deep-sea observatory HAUSGARTEN. *Marine Pollution Bulletin*, 64 (12), 2734–2741.

Best, H., 2009. Structural and Ideological Determinants of Household Waste Recycling: Results from an Empirical Study in Cologne, Germany. *Nature and Culture*, 4 (2), 167–190.

Best, H. and Kneip, T., 2011. The impact of attitudes and behavioral costs on environmental behavior: A natural experiment on household waste recycling. *Social Science Research*, 40 (3), 917–930.

Best, H. and Kroneberg, C., 2012. Die Low-Cost-Hypothese. Theoretische Grundlagen und empirische Implikationen. *Kölner Zeitschrift für Soziologie und Sozialpsychologie*, 64 (3), 535–561.

Blaikie, N.W.H., 1992. The Nature and Origins of Ecological World Views: An Australian Study. *Social Science Quarterly*, 73 (1), 144–65.

Blasius, J. and Gower, J.C., 2005. Multivariate Prediction with Nonlinear Principal Components Analysis: Application. *Quality and Quantity*, 39 (4), 373–390.

Blasius, J. and Greenacre, M., 2006. Correspondence Analysis and Related Methods in Practice. *In*: M. Greenacre and J. Blasius, eds. *Multiple Correspondence Analysis and Related Methods*. London: Chapman & Hall, 4–40.

Blasius, J. and Thiessen, V., 2006. Assessing Data Quality and Construct Comparability in Cross-National Surveys. *European Sociological Review*, 22 (3), 229–242.

Blasius, J. and Thiessen, V., 2012. *Assessing the Quality of Survey Data*. SAGE Publications.

Blind, P.K., 2006. Building trust in government in the twenty-first century: Reviewing of literature and emerging issues. *7th Global Forum on Reinventing Government Building Trust in Government 26–29 June 2007, Vienna, Austria*, 1–31.

Blocker, J.T. and Eckberg, D.L., 1997. Gender and Environmentalism: Results from the 1993 General Social Survey. *Social Science Quarterly*, 78 (4), 841–858.

Blunch, N., 2012. *Introduction to Structural Equation Modeling Using IBM SPSS Statistics and Amos*. SAGE.

Bornschier, V., 2005. Varianten des Kapitalismus in reichen Demokratien beim Übergang in das neue Gesellschaftsmodell. *Kölner Zeitschrift für Soziologie und Sozialpsychologie*, Sonderheft 45, 331–371.

Boudon, R., 2003. Beyond Rational Choice Theory. *Annual Review of Sociology*, 29 (1), 1–21.

Brack, D., 2009. Combating International Environmental Crime. *In*: *Environmental Crime: A Reader*. Portland: Willian Publishing, 485–492.

Brechin, S.R., 1999. Objective Problems, Subjective Values, and Global Environmentalism: Evaluating the Postmaterialist Argument and Challenging a New Explanation. *Social Science Quarterly*, 89, 793–809.

Brechin, S.R. and Kempton, W., 1994. Global Environmentalism: A Challenge to the Post materialism Thesis? *Social Science Quarterly*, 245–269.

Bredl, S., Winker, P., and Kötschau, K., 2012. Survey Methodology – A statistical approach to detect interviewer falsification of survey data. *Survey Methodology*, 8, 1–10.

Buhaug, H., Gleditsch, N., and Theisen, O.M., 2009. Implications of Climate Change for Armed Conflict. *In*: R. Mearns and A. Norton, eds. *Social Dimensions of Climate Change: Equity and Vulnerability in a Warming World*. Washington DC: World Bank, 75–101.

Bundespräsidialamt, 2011. Pressemitteilung 01.08.2011: Bundespräsident fertigt Gesetz zur Änderung des Atomgesetzes aus.

Burr, V., 2003. *Social Constructionism*. Psychology Press.

Buttel, F.H., 1987. New Directions in Environmental Sociology. *Annual Review of Sociology*, 13 (1), 465–488.

Buttel, F.H., 2003. Environmental Sociology and the Explanation of Environmental Reform. *Organization & Environment*, 16 (3), 306–344.

Byrne, B.M., 2009. *Structural Equation Modeling With AMOS: Basic Concepts, Applications, and Programming, Second Edition*. Taylor & Francis.

Catton, W.R. and Dunlap, R.E., 1978. Environmental Sociology: A New Paradigm. *The American Sociologist*, 13, 41–49.

Clark, C., Kotchen, M., and Moore, M., 2003. Internal and external influences on pro-environmental behavior: Participation in a green electricity program. *Journal of Environmental Psychology*, 23 (3), 237–246.

Clinard, M. and Meier, R., 2011. *Sociology of Deviant Behavior*. 14th ed. Wadsworth: Cengage Learning.

Coumou, D. and Rahmstorf, S., 2012. A decade of weather extremes. *Nature Climate Change*, 2 (7), 491–496.

Crepaz, M.M.L., 1995. Explaining national variations of air pollution levels: Political institutions and their impact on environmental policy-making. *Environmental Politics*, 4 (3), 391–414.

Davis, D.W., 2000. Individual Level Examination of Postmaterialism in the U.S.: Political Tolerance, Racial Attitudes, Environmentalism, -aand Participatory Norms. *Political Research Quarterly*, 53 (3), 455–475.

Dekker, P., Ester, P., and Nas, M., 1997. Religion, Culture and Environmental Concern: An Empirical Cross-national Analysis. *Social Compass*, 44 (3), 443–458.

Diekmann, A. and Franzen, A., 1999. The Wealth of Nations and Environmental Concern. *Environment and Behavior*, 31 (4), 540–549.

Diekmann, A. and Meyer, R., 2010. Demokratischer Smog? Eine empirische Untersuchung zum Zusammenhang zwischen Sozialschicht und Umweltbelastungen. *Kölner Zeitschrift für Soziologie und Sozialpsychologie*, 62 (3), 437–457.

Diekmann, A. and Preisendörfer, P., 1998. Environmental Behavior Discrepancies Between Aspirations and Reality. *Rationality and Society*, 10 (1), 79–102.

Diekmann, A. and Preisendörfer, P., 2003. Green and Greenback The Behavioral Effects of Environmental Attitudes in Low-Cost and High-Cost Situations. *Rationality and Society*, 15 (4), 441–472.

Dietz, T., Dan, A., and Shwom, R., 2007. Support for Climate Change Policy: Social Psychological and Social Structural Influences*. *Rural Sociology*, 72 (2), 185–214.

Dietz, T., Fitzgerald, A., and Shwom, R., 2005. Environmental Values. *Annual Review of Environment and Resources*, 30 (1), 335–372.

Dietz, T., Gardner, G.T., Gilligan, J., Stern, P.C., and Vandenbergh, M.P., 2009. Household actions can provide a behavioral wedge to rapidly reduce US carbon emissions. *Proceedings of the National Academy of Sciences*, 106 (44), 18452–18456.

Dietz, T., Rosa, E.A., and York, R., 2012. Environmentally efficient well-being: Is there a Kuznets curve? *Applied Geography*, 32 (1), 21–28.

Dietz, T., Stern, P.C., and Guagnano, G.A., 1998. Social Structural and Social Psychological Bases of Environmental Concern. *Environment and Behavior*, 30 (4), 450–471.

Dinda, S., 2004. Environmental Kuznets Curve Hypothesis: A Survey. *Ecological Economics*, 49 (4), 431–455.

Dinda, S., 2005. A theoretical basis for the environmental Kuznets curve. *Ecological Economics*, 53 (3), 403–413.

Dolnicar, S. and Grün, B., 2009. Environmentally Friendly Behavior Can Heterogeneity Among Individuals and Contexts/ Environments Be Harvested for Improved Sustainable Management? *Environment and Behavior*, 41 (5), 693–714.

Downs, A., 1972. Up and Down with Ecology--The Issue Attention Cycle. *The Public Interest*, 28, 38–50.

Dunlap, R.E., 1989. Public Opinion and environmental Policy. *In*: J.P. Lester, ed. *Environmental Politics and Policy*. Durkham: Duke University Press.

Dunlap, R.E., 2002. Paradigms, Theories, and Environmental Sociology. *In*: *Sociological Theory and the Environment: Classical Foundations, Contemporary Insights*. London: Prentice-Hall, 329–350.

Dunlap, R.E., 2010. The Maturation and Diversification of Environmental Sociology: From Constructivism and Realism to Agnosticism and Pragmatism. *In*: M. Redclift and G. Woodgate, eds. *The International Handbook Of Environmental Sociology*. Cheltenham: Elgar.

Dunlap, R.E. and Catton, W.R., 1979. Environmental Sociology. *Annual Review of Sociology*, 5 (1), 243–273.

Dunlap, R.E. and Catton, W.R., 1994. Struggling with human exemptionalism: The rise, decline and revitalization of environmental sociology. *The American Sociologist*, 25 (1), 5–30.

Dunlap, R.E. and Van Liere, K.D., 1978. *Environmental Concern*. Monticello: Vance.

Dunlap, R.E. and Van Liere, K.D., 2008. The 'New Environmental Paradigm'. *The Journal of Environmental Education*, 40 (1), 19–28.

Dunlap, R.E. and Mertig, A.G., 1995. Global Concern for the Environment: Is Affluence a Prerequisite? *Journal of Social Issues*, 51 (4), 121–137.

Dunlap, R.E. and Mertig, A.G., 1997. Global Environmental Concern: An Anomaly for Postmaterialism. *Social Science Quarterly*, 78 (1), 24–29.

Dunlap, R.E. and York, R., 2008. The Globalization of Environmental Concern and of the Limits of the Postmaterialist Values Explanation: Evidence from Four Multinational Surveys. *Sociological Quarterly*, 49 (3), 529–563.

Duroy, Q.M., 2005. The Determinants of Environmental Awareness and Behavior. *Rensselaer Working papers in economics*, 0501, 1–26.

Elster, J., 1986. *Rational Choice*. New York University Press.

Engel, U. and Pötschke, M., 1998. Willingness to pay for the environment: Social structure, value orientations and environmental behaviour in a multilevel perspective. *Innovation: The European Journal of Social Science Research*, 11 (3), 315–332.

Environmental Investigation Agency, 2008. Environmental Crime: A Threat to the Future. http://www.unodc.org/documents/NGO/EIA_Ecocrime_report_0908_final_draft_low.pdf last access: 15/8/11.

Esping-Andersen, G., 1990. *The Three Worlds of Welfare Capitalism*. Princeton University Press.

European Environment Agency, 2010. The European Environment – State and Outlook 2010: Spain: Freshwater. http://www.eea.europa.eu/soer/countries es/soertopic_view?topic=freshwater last access: 02/04/2013.

Eurostat, 2011. Youth Unemployment. European Commission. http://epp.eurostat.ec.europa.eu statistics_explained/index.php?title=File:Youth_une ployment,_2011Q4_%28%2529.png&filetimestamp=20120502094632 last access: 10/04/2013.

Faris, E., 1937. *The Nature of Human Nature*. New York: McGraw-Hill.

Field, B.C., 1994. *Environmental economics: an introduction*. McGraw-Hill.

Fietkau, H.-J. and Kessel, H., 1981. Umweltlernen. *In*: H.-J. Fietkau and H. Kessel, eds. Königstein: Hain.

Fischbacher, U., Gächter, S., and Fehr, E., 2001. Are people conditionally cooperative? Evidence from a public goods experiment. *Economics Letters*, 71 (3), 397–404.

Foley, J., 2004. Overview. *In*: J. Foley, ed. *Sustainability and Social Justice*. London: Institute of Public Policy Research.

Folmer, H., 2009. Why Sociology is Better Conditioned to Explain Economic Behaviour than Economics. *Kyklos*, 62 (2), 258–274.

Franzen, A., 2003. Environmental Attitudes in International Comparison: An Analysis of the ISSP Surveys 1993 and 2000*. *Social Science Quarterly*, 84 (2), 297–308.

Franzen, A. and Meyer, R., 2004. Klimawandel des Umweltbewusstseins? Eine Analyse des ISSP 2000. *Zeitschrift für Soziologie*, 33 (2), 119–137.

Franzen, A. and Meyer, R., 2010. Environmental Attitudes in Cross-National Perspective: A Multilevel Analysis of the ISSP 1993 and 2000. *European Sociological Review*, 26 (2), 219–234.

Franzen, A. and Vogl, D., 2010. Umweltbewusstsein und der Reichtum von Nationen: Ein Vergleich von WVS, ISSP und EVS. *In*: T. Beckers, K. Birkelbach, J. Hagenah, and U. Rosar, eds. *Komparative empirische Sozialforschung*. VS Verlag für Sozialwissenschaften, 337–362.

Franzen, A. and Vogl, D., 2013a. Time Preferences and Environmental Concern: An Analysis of the Swiss ISSP 2010. *International Journal of Sociology*, 43 (4), 39–62.

Franzen, A. and Vogl, D., 2013b. Two decades of measuring environmental attitudes: A comparative analysis of 33 countries. *Global Environmental Change*, 23 (5), 1001–1008.

Franzen, A., Vogl, D., and Sesterhenn, F., 2009. Gesellschaftliche Reaktion auf die Klimadebatte in Deutschland. *In*: H. Hengstler, ed. *Klima und Energie im Spannungsfeld zwischen Risiko und Verantwortung*. Konstanz: Universität Konstanz, 72–89.

Frey, B.S. and Stutzer, A., 2006. Environmental Morale and Motivation. *University Zurich: Working Papers Series*, 288, 1–27.

Freymeyer, R.H. and Johnson, B.E., 2010. A Cross-Cultural Investigation of Factors Influencing Environmental Actions. *Sociological Spectrum*, 30 (2), 184–195.

Gächter, S., 2007. Conditional cooperation: Behavioral regularities from the lab and the field and their policy implications. *In*: B.S. Frey and A. Stutzer, eds. *Economics and psychology. A Promising New Cross-disciplinary Field*. Cambridge: MIT Press, 19–50.

Gatersleben, B., Steg, L., and Vlek, C., 2002. Measurement and Determinants of Environmentally Significant Consumer Behavior. *Environment and Behavior*, 34 (3), 335–362.

Gelissen, J., 2007. Explaining Popular Support for Environmental Protection A Multilevel Analysis of 50 Nations. *Environment and Behavior*, 39 (3), 392–415.

Gillham, P.F., 2008. Participation in the Environmental Movement Analysis of the European Union. *International Sociology*, 23 (1), 67–93.

Givens, J.E. and Jorgenson, A.K., 2011. The Effects of Affluence, Economic Development, and Environmental Degradation on Environmental Concern: A Multilevel Analysis. *Organization & Environment*, 24 (1), 74–91.

Gould, S.J., 2003. *The Hedgehog, the Fox, and the Magister's Pox: Mending the Gap Between Science and the Humanities*. Three Rivers Press.

Grendstad, G., Stelle, P., Oystein, B., and Stromsnes, K., 2006. *Unique Environmentalism - A Comparative Perspective*. New York: Springer.

Guagnano, G.A., Stern, P.C., and Dietz, T., 1995. Influences on Attitude-Behavior Relationships A Natural Experiment with Curbside Recycling. *Environment and Behavior*, 27 (5), 699–718.

Guha, R., 2000. *Environmentalism: A Global History*. Longman.

Gunningham, N., 2011. Enforcing Environmental Regulation. *Journal of Environmental Law*, 23 (2), 169–201.

Hadler, M. and Haller, M., 2011. Global activism and nationally driven recycling: The influence of world society and national contexts on public and private environmental behavior. *International Sociology*, 26 (3), 315–345.

Halsey, M. and White, R., 2009. Crime, Ecophilosophy and Environmental Harm. *In*: R. White, ed. *Environmental Crime: A Reader*. Portland: Willian Publishing, 27–49.

Hancock, G.R. and Liu, M., 2012. Bootstrapping Standard Errors and Data-Model Fit Statistics in Structural Equation Modeling. *In*: R.H. Hoyle, ed. *Handbook of Structural Equation Modeling*. New York: The Guilford Press, 296–306.

Hardin, G., 1968. The Tragedy of the Commons. *Science*, 162 (3859), 1243–1248.

Hardin, R., 1971. Collective action as an agreeable n-prisoners' dilemma. *Behavioral Science*, 16 (5), 472–481.

Hayes, B.C., 2001. Gender, Scientific Knowledge, and Attitudes toward the Environment: A Cross-National Analysis. *Political Research Quarterly*, 54 (3), 657.

Heberlein, T.A., 2012. *Navigating Environmental Attitudes*. Oxford University Press.

Heinrichs, H. and Gross, M., 2010. Introduction: New Trends and Interdisciplinary Challenges in Environmental Sociology. *In*: M. Gross and H. Heinrichs, eds. *Environmental Sociology*. Springer Netherlands, 1–16.

Hill, M.K., 2004. *Understanding Environmental Pollution: A Primer*. Cambridge University Press.

Hill, M.K., 2010. *Understanding Environmental Pollution*. Cambridge University Press.

Hirschman, A., 1986. Against Parsimony: Three Ways of Complicating Some Categories of Economic Discourse. *In*: A. Hirschman, ed. *Rival Views of Market Society and other Recent Essays*. New York: Vicking, 142–160.

Horbach, J., Blien, U., and Hauff, M. von, 2008. Structural Change and Performance of the German Environmental Sector. *IAB-Discussion Paper*, 36, 1–25.

Hsiang, S.M., Meng, K.C., and Cane, M.A., 2011. Civil conflicts are associated with the global climate. *Nature*, 476 (7361), 438–441.

Hu, L. and Bentler, P.M., 1999. Cutoff criteria for fit indexes in covariance structure analysis: Conventional criteria versus new alternatives. *Structural Equation Modeling: A Multidisciplinary Journal*, 6 (1), 1–55.

Huber, J., 2001. *Allgemeine UmweltSoziologie*. Springer.

Huber, J., 2011. *Allgemeine Umweltsoziologie*. Heidelberg: Springer.

Hunter, L.M., Hatch, A., and Johnson, A., 2004. Cross-National Gender Variation in Environmental Behaviors*. *Social Science Quarterly*, 85 (3), 677–694.

Iacobucci, D., 2010. Structural equations modeling: Fit Indices, sample size, and advanced topics. *Journal of Consumer Psychology*, 20 (1), 90–98.

Ignatow, G., 2006a. Cultural Models of Nature and Society Reconsidering Environmental Attitudes and Concern. *Environment and Behavior*, 38 (4), 441–461.

Ignatow, G., 2006b. Cultural Models of Nature and Society Reconsidering Environmental Attitudes and Concern. *Environment and Behavior*, 38 (4), 441–461.

Inglehart, R., 1995. Public Support for Environmental Protection: Objective Problems and Subjective Values in 43 Societies. *PS: Political Science and Politics*, 28 (1), 57.

Inglehart, R., 1997. *Modernization and Postmodernization: Cultural, Economic, and Political Change in 43 Societies*. Princeton: Princeton University Press.

Inglehart, R.F., 1990. *Culture Shift in Advanced Industrial Society*. Princeton: Princeton University Press.

International Energy Agency (IEA), 2012. Energy Balances of OECD Countries http://www.oecd-ilibrary.org/docserver/download/6112101e.pdf?expires=1364889743&id=id&accname=ocid41014028&checksum=16240F14F3B3EFCE20237AFF98FA250F Last access: 02/04/2013.

Israel, D. and Levinson, A., 2004. Willingness to Pay for Environmental Quality: Testable Empirical Implications of the Growth and Environment Literature. *Contributions in Economic Analysis & Policy*, 3 (1), 1–31.

Jack, B., 2010. Enforcing Member State Compliance with EU Environmental Law: A Critical Evaluation of the Use of Financial Penalties. *Journal of Environmental Law*, eqq023.

Jack, B., 2011. Enforcing Member State Compliance with EU Environmental Law: A Critical Evaluation of the Use of Financial Penalties. *Journal of Environmental Law*, 23 (1), 73–95.

Jacobs, M., 1991. *The green economy: environment, sustainable development, and the politics of the future*. London: Pluto Press.

Jahn, D., 1998. Environmental performance and policy regimes: Explaining variations in 18 OECD-countries. *Policy Sciences*, 31 (2), 107–131.

Jalava, J., 2003. From Norms to Trust The Luhmannian Connections between Trust and System. *European Journal of Social Theory*, 6 (2), 173–190.

Jasper, J.M. and Poulsen, J.D., 1995. Recruiting Strangers and Friends: Moral Shocks and Social Networks in Animal Rights and Anti-Nuclear Protests. *Social Problems*, 42 (4), 493–512.

Jerrett, M., Burnett, R.T., Ma, R., Pope, C.A., 3rd, Krewski, D., Newbold, K.B., Thurston, G., Shi, Y., Finkelstein, N., Calle, E.E., and Thun, M.J., 2005. Spatial analysis of air pollution and mortality in Los Angeles. *Epidemiology*, 16 (6), 727–736.

Jones, R.E. and Dunlap, R.E., 1992. The Social Bases of Environmental Concern: Have They Changed Over Time?1. *Rural Sociology*, 57 (1), 28–47.

Kanagy, C.L., Humphrey, C.R., and Firebaught, G., 1994. Surging environmentalism: Changing public opinion or changing publics? *Social Science Quarterly*, 75 (4), 804–819.

Kellstedt, P.M., Zahran, S., and Vedlitz, A., 2008. Personal efficacy, the information environment, and attitudes toward global warming and climate change in the United States. *Risk analysis: an official publication of the Society for Risk Analysis*, 28 (1), 113–126.

Kemmelmeier, M., Król, G., and Kim, Y.H., 2002. Values, Economics, and Proenvironmental Attitudes in 22 Societies. *Cross-Cultural Research*, 36 (3), 256–285.

Kidd, Q. and Lee, A.-R., 1997. Postmaterialist values and the environment : A critique and reappraisal. Commentaries. *Social science quarterly*, 78 (1), 1–43.

Kijima, M., Nishide, K., and Ohyama, A., 2010. Economic models for the environmental Kuznets curve: A survey. *Journal of Economic Dynamics and Control*, 34 (7), 1187–1201.

Klineberg, S.L., McKeever, M., and Rothenbach, B., 1998. Demographic Predictors of Environmental Concern: It does Make a Difference how it's Measured. *Social Science Quarterly*, 79 (4), 734–753.

Knight, K.W. and Messer, B.L., 2012. Environmental Concern in Cross-National Perspective: The Effects of Affluence, Environmental Degradation, and World Society*. *Social Science Quarterly*, 93 (2), 521–537.

Kollmorgen, R., 2009. PostsozialistischeWohlfahrtsregime in Europa – Teil der „DreiWelten" oder eigener Typus? *In*: B. Pfau-Effinger, S.S. Magdalenić, and C. Wolf, eds. *International vergleichende Sozialforschung*. VS Verlag für Sozialwissenschaften, 65–92.

Kowarik, I., 2011. Novel urban ecosystems, biodiversity, and conservation. *Environmental Pollution*, 159 (8–9), 1974–1983.

Kroneberg, C. and Kalter, F., 2012. Rational Choice Theory and Empirical Research: Methodological and Theoretical Contributions in Europe. *Annual Review of Sociology*, 38 (1), 73–92.

Kuckartz, U., 1997. Grünes Trikot für Deutschland? Das Umweltbewusstsein der Deutschen im internationalen Vergleich. *Zeitschrift für Umweltpolitik & Umweltrecht*, 20 (4), 433–462.

Lachmann, D., forthcoming. Data duplication in cross-national surveys – The Case of the International Social Survey Programme 2010.

Lachmann, D., forthcoming. A Multiple Correspondence Analysis of Two Scales Measuring Environmental Concern in the International Social Survey Programme 2010.

Lappé, F.M., 1988. *Taking Population Seriously*. London: Earthscan.

Lejano, R.P., Tavares-Reager, J., and Berkes, F., 2013. Climate and narrative: Environmental knowledge in everyday life. *Environmental Science & Policy*, 31, 61–70.

Lewis, J.D. and Weigert, A., 1985. Trust as a Social Reality. *Social Forces*, 63 (4), 967.

Liddle, B. and Lung, S., 2010. Age-structure, urbanization, and climate change in developed countries: revisiting STIRPAT for disaggregated population and consumption-related environmental impacts. *Population and Environment*, 31 (5), 317–343.

Liebe, U. and Preisendörfer, P., 2010. Rational Choice Theory and the Environment: Variants, Applications, and New Trends. *In*: M. Gross and H. Heinrichs, eds. *Environmental Sociology*. Springer: Springer Netherlands, 141–157.

Liebe, U., Preisendörfer, P., and Meyerhoff, J., 2011. To Pay or Not to Pay: Competing Theories to Explain Individuals' Willingness to Pay for Public Environmental Goods. *Environment and Behavior*, 43 (1), 106–130.

Liere, K.D.V. and Dunlap, R.E., 1980. The Social Bases of Environmental Concern: A Review of Hypotheses, Explanations and Empirical Evidence. *Public Opinion Quarterly*, 44 (2), 181–197.

Little, T.D., Preacher, K.J., Selig, J.P., and Card, N.A., 2007. New developments in latent variable panel analyses of longitudinal data. *International Journal of Behavioral Development*, 31 (4), 357–365.

Lomborg, B., 2007. *Cool it: The Skeptical Environmentalist's Guide to Global Warming*. Alfred A. Knopf.

Lomborg, B., 2010. *Smart Solutions to Climate Change: Comparing Costs and Benefits*. Cambridge: Cambridge University Press.

Longhofer, W. and Schofer, E., 2010. National and Global Origins of Environmental Association. *American Sociological Review*, 75 (4), 505–533.

Luhmann, N., 1979. *Trust and Power: Two Works*. New York: Wiley.

Mafessanti, M., 2009. Responsibility for Environmental Damage under International Law: Can MNCs Bear the Burden… and How. *Buffalo Environmental Law Journal*, 17, 87.

Mahoney, W., 2010. *The History of the Czech Republic and Slovakia*. ABC-CLIO.

Maloney, M.P. and Ward, M.P., 1973. Ecology: Lets hear from the people: An objective scale for the measurement of ecological attitudes and knowledge. *American Psychologist*, 28 (7), 583–586.

Maloney, M.P., Ward, M.P., and Braucht, N.G., 1975. A revised scale for the measurement of ecological attitudes and knowledge. *American Psychologist*, 30 (7), 787–790.

Mares, D., 2010. Criminalizing Ecological Harm: Crimes Against Carrying Capacity and the Criminalization of Eco-Sinners. *Critical Criminology*, 18 (4), 279–293.

Matthews, M.M., 2001. Cleaning Up Their Acts: Shifts of Environment and Energy Policies in Pluralist and Corporatist States. *Policy Studies Journal*, 29 (3), 478–498.

McAdam, D., 1986. Recruitment to High-Risk Activism: The Case of Freedom Summer. *The American Journal of Sociology*, 92 (1), 64–94.

McPherson, M.A. and Nieswiadomy, M.L., 2005. Environmental Kuznets curve: threatened species and spatial effects. *Ecological Economics*, 55 (3), 395–407.

Meadows, D.H., Meadows, D.L., Randers, J., and Behrens III, W.W., 1972. *The Limits to growth: a report for the Club of Rome's project on the predicament of mankind*. Universe Books.

Mearns, R. and Norton, A., 2009. Equity and Vulnerability in a Warming World: Introduction and Overview. *In*: R. Mearns and A. Norton, eds. *Social Dimensions of Climate Change: Equity and Vulnerability in a Warming World*. Washington DC: World Bank, 1–44.

Merton, R.K., 1936. The Unanticipated Consequences of Purposive Social Action. *American Sociological Review*, 1 (6), 894.

Messer, C.M. and Shriver, T.E., 2009. Corporate Responses to Claims of Environmental Misconduct: The Case of Phelps Dodge and Blackwell, Oklahoma. *Deviant Behavior*, 30 (7), 647–668.

Meyer, R. and Liebe, U., 2010. Are the affluent prepared to pay for the planet? Explaining willingness to pay for public and quasi-private environmental goods in Switzerland. *Population and Environment*, 32 (1), 42–65.

Misztal, B.A., 2001. Trust and cooperation: the democratic public sphere. *Journal of Sociology*, 37 (4), 371–386.

Mohai, P. and Saha, R., 2006. Reassessing racial and socioeconomic disparities in environmental justice research. *Demography*, 43 (2), 383–399.

Nawrotzki, R.J., 2012. The Politics of Environmental Concern A Cross-National Analysis. *Organization & Environment*, 25 (3), 286–307.

Nepstad, S.E. and Smith, C., 1999. Rethinking Recruitment to High-Risk/Cost Activism: The Case of the Nicaragua Exchange. *Mobilization: An International Quarterly*, 4 (1), 25–40.

Neumayer, E., 2003. Are left-wing party strength and corporatism good for the environment? Evidence from panel analysis of air pollution in OECD countries. *Ecological Economics*, 45 (2), 203–220.

New Zealand Government, 2013. New Zealand Economic and Financial Overview 2013. http://www.treasury.govt.nz/economy/overview/2013/nzefo-13.pdf Last access: 02/04/2013.

Nisbet, E.K., Zelenski, J.M., and Murphy, S.A., 2009. The Nature Relatedness Scale Linking Individuals' Connection With Nature to Environmental Concern and Behavior. *Environment and Behavior*, 41 (5), 715–740.

Nolan, J.M., 2010. 'An Inconvenient Truth' Increases Knowledge, Concern, and Willingness to Reduce Greenhouse Gases. *Environment and Behavior*, 42 (5), 643–658.

Norwegian Water Resources and Energy Directorate, 2011. Energy in Norway. http://www.nve.no/Global/Energi/Analyser/Energi%20i%20Norge%20folder/FOLDE2011.pdf Last access: 01/04/2013.

Nuclear Accident Independent Investigation Commission (NAIIC), 2012. The Official Report of the Fukushima Nuclear Accident Independent Investigation Commission: Executive Summary. The National Diet of Japan. http://naiic.go.jp/wp-content/uploads/2012/07/naiic_report_lo_res.pdf last access: 05/07/2012.

Obach, B.K., 2009. Labor and the Environment. *In*: K.A. Gould, ed. *Twenty Lessons in Environmental Sociology*. New York: Oxford University Press, 228–243.

Olofsson, A. and Öhman, S., 2006. General Beliefs and Environmental Concern Transatlantic Comparisons. *Environment and Behavior*, 38 (6), 768–790.

Olson, M., 1965. *The Logic of Collective Action*. Cambridge: Harvard University Press.

Olsson, S., 1994. What Happened to the 'Silent Revolution'? Postmaterialist Values Across Generations. *In*: M. Kloep, ed. *Youth in a Changing World*. Östersund: Mid Sweden University.

Opp, K.-D., 1999. Contending Conceptions of the Theory of Rational Action. *Journal of Theoretical Politics*, 11 (2), 171–202.

Ostrom, E., 2000. Collective Action and the Evolution of Social Norms. *Journal of Economic Perspectives*, 14 (3), 137–158.

Ostrom, P.D.E., 2011. Handeln statt Warten: Ein mehrstufiger Ansatz zur Bewältigung des Klimaproblems. *Leviathan*, 39 (2), 267–278.

Pampel, F.C. and Hunter, L.M., 2012. Cohort Change, Diffusion, and Support for Environmental Spending in the United States[1]. *American Journal of Sociology*, 118 (2), 420–448.

Pelletier, L.G., Tuson, K.M., and Haddad, N.K., 1997. Client Motivation for Therapy Scale: a measure of intrinsic motivation, extrinsic motivation, and amotivation for therapy. *Journal of personality assessment*, 68 (2), 414–35.

Pepper, M., Jackson, T., and Uzzell, D., 2011. An Examination of Christianity and Socially Conscious and Frugal Consumer Behaviors. *Environment and Behavior*, 43 (2), 274–290.

Perk, M. van der, 2006. *Soil and Water Contamination: From Molecular to Catchment Scale*. CRC Press.

Poortinga, W., Steg, L., Vlek, C., and Wiersma, G., 2003. Household preferences for energy-saving measures: A conjoint analysis. *Journal of Economic Psychology*, 24 (1), 49–64.

Porter, M.E. and Linde, C. van der, 1995. Toward a New Conception of the Environment-Competitiveness Relationship. *Journal of Economic Perspectives*, 9 (4), 97–118.

Preisendörfer, P., 1996a. *Soziologie und Ökologie I: Grundlagen und Problemstellungen der Umweltsoziologie*. Rostrock: Universität Rostock.

Preisendörfer, P., 1996b. Ökologisches Bewusstsein in Ost- und Westdeutschland. *Zeitschrift für Umweltpolitik & Umweltrecht*, 19 (1), 1–20.

Raleigh, C. and Jordan, L., 2009. Climate Change and Migration: Emerging Patterns in the Developing World. *In*: R. Mearns and A. Norton, eds. *Social Dimensions of Climate Change: Equity and Vulnerability in a Warming World*. Washington DC: World Bank, 103–131.

Räthzel, N. and Uzzell, D., 2011. Trade unions and climate change: The jobs versus environment dilemma. *Global Environmental Change*, 21 (4), 1215–1223.

Rees, W.E., 1992. Ecological footprints and appropriated carrying capacity: what urban economics leaves out. *Environment and Urbanization*, 4 (2), 121–130.

Ribot, J., 2009. Vulnerability Does Not Fall from the Sky: Toward Multiscale, Pro-poor Climate Policy. *In*: R. Mearns and A. Norton, eds. *Social Dimensions of Climate Change: Equity and Vulnerability in a Warming World*. Washington DC: World Bank, 47–74.

Ringquist, E.J., 2011. Trading Equity for Efficiency in Environmental Protection? Environmental Justice Effects from the SO2 Allowance Trading Program*. *Social Science Quarterly*, 92 (2), 297–323.

Rogers, E.M., 2003. *Diffusion of Innovations, 5th Edition*. New York: Simon and Schuster.

Røpke, I., 2009. Theories of practice — New inspiration for ecological economic studies on consumption. *Ecological Economics*, 68 (10), 2490–2497.

Rössel, J., 2006. Ronald Inglehart: Daten auf der Suche nach einer Theorie — Analysen des weltweiten Wertewandels. *In*: S. Moebius and D. Quadflieg, eds. *Kultur. Theorien der Gegenwart*. VS Verlag für Sozialwissenschaften, 545–556.

Samuelson, P.A., 1954. The Pure Theory of Public Expenditure. *The Review of Economics and Statistics*, 36 (4), 387.

Schäfer, C., Schräpler, J.-P., Müller, K.-R., and Wagner, G.G., 2005. Automatic Identification of Faked and Fraudulent Interviews in the German SOEP. *Schmollers Jahrbuch*, 125, 183–193.

Schan, J., 2000. Muss die Low-Cost-Hypothese modifiziert werden? *Umweltpsychologie*, 4 (3), 100–116.

Schlosberg, D., 2007. *Defining Environmental Justice: Theories, Movements, and Nature*. OUP Oxford.

Schnaiberg, A., 1980. *The environment, from surplus to scarcity*. New York: Oxford University Press.

Schor, J.B., 2005. Prices and quantities: Unsustainable consumption and the global economy. *Ecological Economics*, 55 (3), 309–320.

Schubert, J., Wolbring, T., and Gill, B., 2013. Settlement Structures and Carbon Emissions in Germany: The Effects of Social and Physical Concentration on Carbon Emissions in Rural and Urban Residential Areas. *Environmental Policy and Governance*, 23 (1), 13–29.

Schuh, H., 2012. Blutiger Ökostrom [Bloody Green Electricity]. *Die Zeit*, 12 Jan, p. 29.

Schultz, P.W., Gouveia, V.V., Cameron, L.D., Tankha, G., Schmuck, P., and Franěk, M., 2005. Values and their Relationship to Environmental Concern and Conservation Behavior. *Journal of Cross-Cultural Psychology*, 36 (4), 457–475.

Schulz, W.P., 2002. Knowledge, information, and household recycling: Examining the knowledge-deficit model of behaviour change. *In*: T. Dietz and P.C. Stern, eds. *New tool for environmental protection: Education, information, and voluntary measures*. Washington DC: The National Academic Press, 67–82.

Scruggs, L., 2001. Is There Really a Link Between Neo-Corporatism and Environmental Performance? Updated Evidence and New Data for the 1980s and 1990s. *British Journal of Political Science*, 31 (04), 686–692.

Scruggs, L.A., 1999. Institutions and Environmental Performance in Seventeen Western Democracies. *British Journal of Political Science*, 29 (01), 1–31.

Seligman, A.B., 1997. *The Problem of Trust*. Princeton: Princeton University Press.

Sen, A., 1982. *Poverty and Famines: An Essay on Entitlement and Deprivation*. Oxford: OUP Oxford.

Shepard, P.M., Northridge, M.E., Prakash, S., and Stover, G., 2002. Advancing Environmental Justice through Community-Based Participatory Research. *Environmental Health Perspectives*, 110 (2), 139–140.

Sherman, J.D. and Smith, H.L., 1984. THE INFLUENCE OF ORGANIZATIONAL STRUCTURE ON INTRINSIC VERSUS EXTRINSIC MOTIVATION. *Academy of Management Journal*, 27 (4), 877–885.

Sinha, R.K. and Singh, S., 2010. *Green plants and pollution: Nature's Technology for Abating and Combating Environmental Pollution*. New York: Nova.

Skarbek-Kozietulska, A., Preisendörfer, P., and Wolter, F., 2012. Leugnen oder gestehen? Bestimmungsfaktoren wahrer Antworten in Befragungen. *Zeitschrift für Soziologie*, 41 (1), 5–23.

Stauffacher, M., 2010. Beyond Neocorporatism? Transdisciplinary Case Studies as a Means for Collaborative Learning in Sustainable Development. *In*: M. Gross and H. Heinrichs, eds. *Environmental Sociology*. Dodrecht: Springer, 201–216.

Steffen, W., Persson, A., Deutsch, L., Zalasiewicz, J., Williams, M., Richardson, K., Crumley, C., Crutzen, P., Folke, C., Gordon, L., Molina, M., Ramanathan, V., Rockstrom, J., Scheffer, M., Schellnhuber, H.J., and Svedin, U., 2011. The Anthropocene: From Global Change to Planetary Stewardship. *Ambio*, 40 (7), 739–761.

Stern, P.C., 2000. New Environmental Theories: Toward a Coherent Theory of Environmentally Significant Behavior. *Journal of Social Issues*, 56 (3), 407–424.

Swami, V., Chamorro-Premuzic, T., Snelgar, R., and Furnham, A., 2011. Personality, individual differences, and demographic antecedents of self-reported household waste management behaviours. *Journal of Environmental Psychology*, 31 (1), 21–26.

Tabernero, C. and Hernández, B., 2011. Self-Efficacy and Intrinsic Motivation Guiding Environmental Behavior. *Environment and Behavior*, 43 (5), 658–675.

Tobler, C., Visschers, V.H.M., and Siegrist, M., 2011. Organic Tomatoes Versus Canned Beans How Do Consumers Assess the Environmental Friendliness of Vegetables? *Environment and Behavior*, 43 (5), 591–611.

Tobler, C., Visschers, V.H.M., and Siegrist, M., 2012. Addressing climate change: Determinants of consumers' willingness to act and to support policy measures. *Journal of Environmental Psychology*, 32 (3), 197–207.

Tomkins, K., 2004. Police, Law Enforcement and the Environment. *Current Issues in Criminal Justice*, 16, 294.

Torgler, B. and García-Valiñas, M.A., 2007. The determinants of individuals' attitudes towards preventing environmental damage. *Ecological Economics*, 63 (2–3), 536–552.

Traxler, F., Blaschke, S., and Kittel, B., 2001. *National Labour Relations in Internationalized Markets: A Comparative Study of Institutions, Change, and Performance*. Oxford: Oxford University Press.

U.S. National Climate Data Centre, 2011. Global Analysis - March 2014 | State of the Climate [online]. Available from: http://www.ncdc.noaa.gov/sotc/global/ [Accessed 19 Mar 2011].

Umweltbundesamt (Federal Environment Agency), 2010. Environmental Laws / Association Claims. Project Environmental Protection. http://www.umweltbundesamt.de/umweltrecht-e/umweltgesetzbuch.htm last access: 12/08/11.

United Nations, 2010. *World Urbanization Prospects: The 2009 Revision. Highlights*. New York: United Nations.

Uslaner, E.M., 2002. *The Moral Foundations of Trust*. Cambridge University Press.

Vandenbergh, M.P. and Steinemann, A.C., 2007. Carbon-Neutral Individual, The. *New York University Law Review*, 82, 1673–1741.

Vlahov, D. and Galea, S., 2002. Urbanization, urbanicity, and health. *Journal of urban health: bulletin of the New York Academy of Medicine*, 79 (4 Suppl 1), S1–S12.

Vrooman, J.C., 2012. Regimes and cultures of social security: Comparing institutional models through nonlinear PCA. *International Journal of Comparative Sociology*, 53 (5–6), 444–477.

Walter, S., 2010. Trust and Cooperation as Requirements for Maintaining Environmental Governance Capacity. *In*: M. Gross and H. Heinrichs, eds. *Environmental Sociology*. Springer Netherlands, 123–139.

Walters, R., 2010. Toxic Atmospheres Air Pollution, Trade and the Politics of Regulation. *Critical Criminology*, 18 (4), 307–323.

Walton, T. and Austin, D.M., 2011. Pro-Environmental Behavior in an Urban Social Structural Context. *Sociological Spectrum*, 31 (3), 260–287.

Wang, Y., Kang, L., Wu, X., and Xiao, Y., 2013. Estimating the environmental Kuznets curve for ecological footprint at the global level: A spatial econometric approach. *Ecological Indicators*, 34, 15–21.

Weber, M., 1978. *Economy and Society: An Outline of Interpretive Sociology*. Berkeley: University of California Press.

Welte, T.H.L. and Anastasio, P.A., 2010. To Conserve or Not to Conserve: Is Status the Question? *Environment and Behavior*, 42 (6), 845–863.

White, R., 2009. Environmental Issues and the Criminological Imagination. *In*: R. White, ed. *Environmental Crime: A Reader*. Portland: Willian Publishing, 63–83.

White, R., 2011. *Transnational Environmental Crime: Toward an Eco-global Criminology*. Routledge.

Winter, D.D.N., 2000. Some big ideas for some big problems. *American Psychologist*, 55 (5), 516–522.

Witzke, H.P. and Urfei, G., 2001. Willingness To Pay for Environmental Protection in Germany: Coping With the Regional Dimension. *Regional Studies*, 35 (3), 207–214.

World Bank, 2013. GDP growth (annual %). http://data.worldbank.org/indicator/NY.GDP.MKTP.KD.ZG/countries last access: 14/03/2013.

Wray-Lake, L., Flanagan, C.A., and Osgood, D.W., 2010. Examining Trends in Adolescent Environmental Attitudes, Beliefs, and Behaviors Across Three Decades. *Environment and Behavior*, 42 (1), 61–85.

York, R. and Clark, B., 2010. Critical Materialism: Science, Technology, and Environmental Sustainability*. *Sociological Inquiry*, 80 (3), 475–499.

Zitzer, S.E., 2009. Renewable Energy Policy and Wind Energy Development in Germany. *UFZ-Diskussionspapiere*, 8/2009, 1–36.

Appendix

Table A 1: Participating countries ISSP 2010

	N	%
Argentina	1130	2,5
Austria	1019	2,3
Belgium - Flanders	1142	2,5
Bulgaria	1003	2,2
Canada	985	2,2
Chile	1436	3,2
Croatia	1210	2,7
Czech Republic	1428	3,2
Denmark	1305	2,9
Finland	1211	2,7
France	2253	5,0
Germany-East	418	,9
Germany-West	989	2,2
Great Britain	928	2,1
Israel-Arabs	193	,4
Israel-Jews	1023	2,3
Japan	1307	2,9
Korea (South)	1576	3,5
Latvia	1000	2,2
Lithuania	1023	2,3
Mexico	1637	3,6
New Zealand	1172	2,6
Norway	1382	3,1
Philippines	1200	2,7
Russia	1619	3,6
Slovak Republic	1159	2,6
Slovenia	1082	2,4
South Africa	3112	6,9
Spain	2560	5,7
Sweden	1181	2,6
Switzerland	1212	2,7
Taiwan	2209	4,9
Turkey	1665	3,7
United States	1430	3,2
Total	45199	100,0

Table A 2: The evaluative dimension of environmental concern

	Items 1	Item 2	Item 3
Austria	V36	V35	V34
Belgium - Flanders	V32	V23	V34
Canada	V23	V36	V25
Croatia	V35	V36	V34
Czech Republic	V34	V35	V36
Denmark	V36	V25	V23
Finland	V36	V25	V35
France	V23	V36	V34
Germany East	V35	V23	V34
Germany West	V23	V35	V32
Great Britain	V23	V36	V25
Japan	V36	V32	V25
New Zealand	V23	V25	V36
Norway	V36	V23	V25
Slovenia	V35	V32	V23
Spain	V36	V23	V32
Sweden	V36	V23	V34
Switzerland	V36	V23	V35
USA	V23	V36	V25

www.ingramcontent.com/pod-product-compliance
Ingram Content Group UK Ltd.
Pitfield, Milton Keynes, MK11 3LW, UK
UKHW021841100926
8556IPUK00019B/990

9 783631 673478